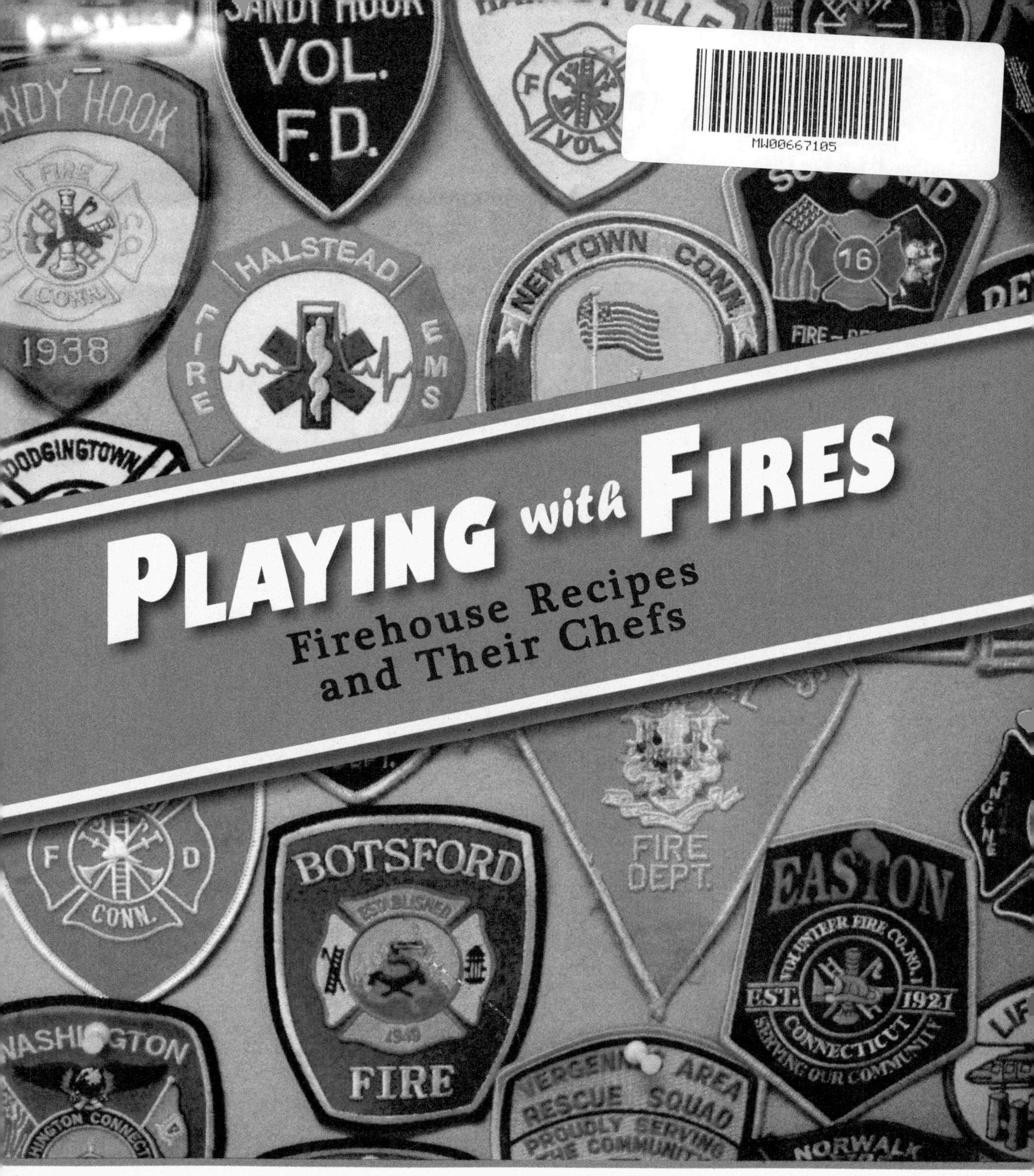

PLAYING with FIRES

Firehouse Recipes and Their Chefs

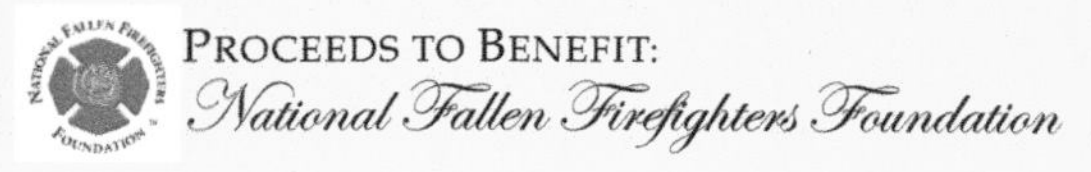

PROCEEDS TO BENEFIT:
National Fallen Firefighters Foundation

STEVEN W. SILER

From Steven: There are four people that I have to recognize...thank you for your belief in me
Dr. Michael Sullivan, Medical Director-San Juan County
Chief Mike Harris, Orcas Fire and Rescue
Division Chief Patrick Shepler, Orcas Fire and Rescue
Division Chief Jim Peeples, South Whatcom Fire and Rescue

And to the Charleston 9. You will not be forgotten. I miss you, Frenchie.
Engine 15 Captain Louis Mulke
Engine 16 Captain Mike Benke
Engine 16 Firefighter Melven Champaign
Engine 19 Captain William "Billy" Hutchinson
Engine 19 Engineer Bradford "Brad" Baity
Engine 19 Firefighter James "Earl" Drayton
Tower 5 Engineer Mark Kelsey
Tower 5 Engineer Michael French
Tower 5 Firefighter Brandon Thompson

Life of a Firefighter from the incomparable Master of the Voodoo Hoodoo, Chief Shepl

Layout by Steven W. Siler

Photography by the fire departments themselves and publicly available sources

Printed in the United States of America

This book is dedicated to the emergency responders...

From the first frantic call to 9-1-1
To the comforting hands at the Emergency Deprtment
You give your time...

away from spouses,
away from friends,
away from children,
And yes, even from meals...

To assure all of us:

"Tonight, I will make it better for you
no matter what,
I will watch over you..."

Proceeds from the sale of every book will benefit the
National Fallen Firefighters Foundation

I have always wondered if anyone really reads the Table of Contents. Now since this is a cookbook, I should have organized everything under its proper heading, like soups, pasta, desserts and the like. This is not just a cookbook as much as a Culinary Postcard; a celebration of the Fire Departments themselves...about the places, history, apparatus and of course, the people.

Four firefighters enter the kitchen as thick black smoke pours from the stove. As the acrid smoke rolls slowly across the ceiling, they determine that the fire is, "contained to the oven" and that it is safe to remove the burned food. They remove a pot roast that will have to be, "identified by dental records."

The meal is ruined; but the firehouse is safe! Funny as this confession is; it is true. In fact, a few fire stations have not been so lucky! Modern fire stations actually have a safety feature that turns off the stove when the Alarm is triggered by the local 911 Dispatch Center.

When food goes into the oven at a firehouse, emergency "Calls" happen. Surely there is no real relationship, but ask any firefighter if meals cause emergency calls and they will usually say, "Yes", They will often have a horror story to back it up. We are a superstitious group, but with good reason. If our shift has been slow and someone makes the offhand comment, "slow shift"; the alarm is sure to sound.

Calls to a fire station come in all forms. 911 calls are transferred to the nearest fire station within seconds, but some people call the local firehouse telephone instead; or drive-up with a baby to "Drop off" or a spouse having a heart attack. Or a geriatric Golden Retriever with a marrow bone stuck on his lower jaw.

While urban departments are often staffed with full-time paid firefighters, most fire departments are staffed by volunteers. But whether volunteer or paid, food preparation and consumption is a meaningful part of fire house culture.

The kitchen is sometimes called, "The Beanery"; but not because of coffee beans. Although coffee is one of the four food groups, the Beanery is named for Beans, baby; legumes that cost next to nothing, simmer all day long, and can feed an army of broke hungry heroes.

For firefighters, emergency medical technicians (EMTs) and Paramedics, their number one job is responding to calls. Their second most important job is being ready to respond to calls. That

requires constant study and working with the fire engines and specialty vehicles, including ambulances. It requires study, care and maintenance of all the tools and equipment they carry. It requires being physically and mentally fit to, "Respond" to whatever the next alarm may call upon them to do.

Firefighters work in, "the worst day of your life business". A few hours after my friends almost burned their own firehouse down, they responded to an upstairs apartment fire. The crew "split", and while two entered the apartment with a hose line, the other two positioned a ladder at the back bedroom window. Walking through a living room, "fully involved in fire", Lieutenant James G. found a four year-old girl under the bed and handed her out the window to a fellow crewmember on the ladder.

Firefighters are trusted. What other job involves being invited into a stranger's home, during their greatest personal emergency, no questions asked? While the shrill sound of the Alarm bell often spells a scene of chaos; firefighters use the Firehouse and the firehouse Routine as a Sanctuary, and an Antidote to that chaos. One Routine is the comfort that for a few bucks and the skill of the cook, you can count on a great group meal. When we sit down to eat a meal we use this time together as the perfect venue to discuss, debrief and defuse the last call, and be ready for the next.

We hope you enjoy and share these firehouse recipes and the great meals they create. We are grateful you have chosen the book and hope you take a moment to remember as you sit down to breakfast, lunch or dinner, that a firefighter might be jumping up from their meal at a firehouse, or from home, to answer a call for help from a stranger having, "the worst day of their life."

Patrick Shepler,
Firefighter/Paramedic
Captain, Orcas Island Fire Rescue
Orcas Island, Washington State

"If Prometheus was worthy of the wrath of heaven for kindling the first fire upon earth, how ought all the gods honor the men who make it their professional business to put it out?"
John Godfrey Saxe

SLOAN'S JAMBALAYA

The Mobile Fire-Rescue Department became a Paid Professional Fire Department September 1, 1888, but to better understand the history of the department we need to take a brief look at the years preceding 1888. Six years after President Madison ordered Mobile taken from Spanish West Florida to become part of the Mississippi Territory, Mobile began to organize itself to try to protect life and property. Creole Steam Fire Company No. 1 is believed to be the first Volunteer Fire Company followed closely by Neptune Engine Fire Company No. 2.

2 (12oz) cans beef consomme
2 (12oz) cans French onion soup
2 (8oz) cans tomato paste
2 cans Rotel tomatoes
½ C. chopped parsley
½ C. chopped green onion
2 diced bell peppers
6 bay leaves
2 Tbsp. thyme
4 C. converted rice
½ lb. salted whole butter
2 lbs. smoked sausage

1 whole chicken
1 halved onion
2 Tbsp. black pepper
2 Tbsp. salt
2 bay leaves

1. Combine top ingredients in a large cast iron Dutch oven. Bring to a boil stirring frequently and place in oven at 350 for 2 hours.

2. In a large pot cover chicken with water and add remaining ingredients. Bring to a boil and cook until done. Allow chicken to cool and separate.

3. Add chicken to Dutch oven, stir thoroughly, cover and place back in oven for remaining time.

Serves 6-10 firemen.

Fire Houses used to have spiral staircases to prevent the horses from climbing the stairs. The pole was so the FF could easily get down stairs when the bell rang for a fire.

Smoked Tuna Dip/Spread

The area that now comprises Orange Beach was first settled in the mid-1860s, with the western portion of the present town being known as Orange Beach, the central portion being known as Caswell, and the eastern end being known as Bear Point. This latter section of the present-day island was a Creek settlement until the early nineteenth century. The Orange Beach Volunteer Fire Department was founded in 1961. Our motto now is "Orange Beach Fire Rescue…Protecting Paradise."

3 cans of white albacore tuna-packed in water (drained well)
(1) 8 oz package of fat free cream cheese
1-2 tsp. of liquid smoke flavoring
2 chopped green onions
salt to taste
¼ tsp. lemon pepper
½ tsp. garlic powder (not garlic salt)
¼ tsp. of shrimp boil seasoning

1. Let the cream cheese soften to room temp.

2. Mix in the drained tuna and remaining ingredients.

3. Serve with crackers, chips, pita chips, or as a wrap with lettuce and tomato.

Orange Beach Fire Department
Firefighter Suzanne Moeller, Orange Beach, Alabama

Before the invention of the horse drawn fire engine, fires were controlled by a "bucket brigade". A line of people passing water in buckets down a line to extinguish a fire and then back again to refill the buckets.

MSA

Beer Dip

Orange Beach Fire/Rescue operates two manned stations on a 24/48 schedule with 3 shifts of eleven firefighters each. All manned apparatus are ALS with at least one paramedic assigned at all times.

In addition, all multi company incidents are overseen by the Battalion Chief on duty. Battalion Chiefs work the same 24/48 schedule as the firefighters they supervise.

8 oz package fat free cream cheese
16 oz package of shredded cheese-any kind (I like sharp cheddar)
1 individual pkg. of Hidden Valley Ranch mix
12 oz cold beer, any brand (dark beer/flavored beer is not recommended-but Blue Moon and Stella Artois are great)

1. Mix top three ingredients well.

2. Pour enough beer to thin down the ingredients to your liking. It will look crazy at first, but the more you mix, the more smooth it will become.

3. Refrigerate for about 20 minutes.

4. Serve with crackers, chips, veggies, or French bread.

Starting them young on learning CPR

Firefighter Suzanne Moeller, Orange Beach, Alabama

A "pipe man" or "nozzle man" is a firefighter whose job is to put water on the fire or the wet stuff on the red stuff.

Suzy's Famous Meatloaf

Every year, Orange Beach Fire and Rescue holds an annual Barbeque Fundraiser. which has proven to be very popular with local community. Smoked Boston butts are transformed into delicious pulled pork, and all for charity.

3 lbs. lean ground beef
1½ C. Italian Bread crumbs
2 eggs
¼ C. Worcestershire sauce
¼ C. A1 steak sauce
1 pkg. of Lipton Onion Soup Mix-dry
2 chopped green onions

Topping:
2 C. ketchup
1 C. brown sugar

1. Heat oven to 350.

2. Mix all ingredients (excluding topping) together well with your HANDS (be sure eggs are mixed in well).

3. Form into a football shape about 2-3 inches thick and place on a broiler pan lined with foil (so the grease drips to the bottom pan).

4. Bake uncovered for about an hour and 20 minutes.

5. Mix the ketchup and brown sugar together well and pour over the loaf. Bake for an additional 20 minutes.

6. Take out and let rest for at least 15 minutes (making sure its done in the middle).

7. Best served with potatoes and green beans!

Firefighter Suzanne Moeller, Orange Beach, Alabama

Back in the early days when cars first appeared on the roads, they were all black. So to help clear the roads when a fire truck was coming through, they painted them red so they would stand out.

CHAR'S PASTA SALAD

The Central Mat-Su Fire Department (CMSFD) is also known as Mat-Su Borough Emergency Services District One. CMSFD is the second largest and second busiest fire department in Alaska, after Anchorage. The department operates a combination system of full-time staff, supplemented with paid-on-call responders. There are three sections within the department, EMS, fire, and rescue; each with its own funding, membership requirements, and equipment. The department currently has over 125 members which includes 12 full-time staff, 4 explorers, and 7 auxiliary volunteers.

tri-color pasta (cooked)
1 purple onion finely chopped
1 C. feta cheese crumbled or diced
1 or 2 red and/or yellow pepper finely chopped
1 C. radish finely chopped
1 packet good seasons Italian dressing (made)
1 can black olives sliced
2 C. broccoli and/or cauliflower pieces

1. Mix veggies together then add pasta, dressing and feta.

Volunteer firefighters cover more square miles, but career firefighters serve more population. There are only 30,000 fire departments in the United States and most of them are volunteer departments.

HOLIDAY POTATOES

The Wasilla Volunteer Fire Department (WFD) was originally organized in March of 1960, shortly after Alaska statehood in 1959. It was the third fire department in the Mat-Su Valley, the city of Palmer and the Butte area fire departments had been in operation for 3 to 4 years before this. The first fire chief was Jake Wright, a veteran of WWII who worked for the Alaska Railroad in Wasilla.

4 lbs. unpeeled potatoes cooked and drained
1 C. chopped onion
¼ butter
1 can cream of celery soup
1 pt. sour cream
1½ C. shredded chedder cheese
½ C. bread crumbs
3 Tbsp. melted butter

1. Remove skin from potatoes, then shred into bowl.

2. Saute onions in butter until tender.

3. Remove from heat and stir in soup and sour cream.

4. Pour over potatoes and chees, mix well.

5. Turn into greased pan,cover and refrigerate overnight.

6. Before baking, sprinkle with bread crumbs and drizzle with melted butter.

7. Bake at 350 for 1 hour or until golden brown.

Firefighter Char Avril, Alaska

Our family loves this dish served with ham at Christmas time.

Enchiladas

Kingman, Arizona, was founded in 1882, when Arizona was only Arizona Territory. Situated in the Hualapai Valley between the Cerbat and Hualapai mountain ranges, Kingman is known for its very modest beginnings as a simple railroad siding. The city of Kingman was named for Lewis Kingman, who surveyed along the Atlantic and Pacific Railroad's right-of-way

2 or 3 medium chicken breasts
diced onion
light butter or oil
salt to taste
pepper to taste
1 can cream of mushroom soup
1 can cream of chicken soup
1 can chicken broth
1 small can of diced green peppers
1 small can of diced jalapenos

1. Brown 2 or 3 medium chicken breasts with diced onion and light butter or oil.
2. Season to preference. I usually just add a touch of salt and pepper.
3. Drain any excess oil or butter; cut into cubes, (or shred/pull) about ½ to ¾" in size.
4. In same pan return cubed chicken; add 1 can cream of mushroom soup, 1 can cream of chicken soup, 1 can chicken broth, 1 small can of diced green peppers, 1 small can of diced jalapenos. (Caution on last two ingredients. Sometimes better to start out easy and add to desired heat level.)
5. Simmer on stove top until you can't wait any longer. Or you believe ingredients have sufficiently exchanged flavors.
6. Grated cheese can be added as a thickening agent while simmering.
7. Pour on to medium tortilla shells, (flour or corn) with cheese and layer like lasagna.
8. Heat in oven on serving plate.
9. On removal, add salsa and or sour cream to taste. Serve with Tostitos or Doritos. Also makes a great game day dip.

CAUTION:
Heat builds up when refrigerated. Great for leftovers.

Kingman Fire Department
Firefighter Mac Nelson, Kingman, Arizona

An average of 100 firefighters a year die in the line of duty. Referred to a LODD, Line-of-Duty Death. The number one reason...heart attacks.

Scottish Tacos

The Kingman Fire Department is a part of the City of Kingman municipal government. The department consists of career and volunteer employees providing prevention and response activities. The current personnel roster consists of 12 part-time and 48 full-time career personnel, 3 civilian support personnel, 2 hydrant maintenance personnel.

1 lb. ground beef
4-6 medium potatoes
1 can of whole kernel corn
2-3 green onions
4 large flour burrito shells
grated cheddar cheese (Mexican mix substitute)
A-1 steak sauce (Scottish salsa)

1. Brown burger in 6 to 8 quart pot with chopped green onion.

2. Brown diced potatoes in frying pan. I prefer about half inch.

3.Drain corn.

4. Drain fat from burger.

5. Add potatoes and drained corn.

6. Simmer on stove top till flavors mix and potatoes are soft.

7. Heat/brown burrito shells.

8. Sprinkle cheese to desired amounts.

9. Add taco mix.

10. Add light amounts of A-1 at first. Don't over do it unless you are a fanatic for A-1.

Kingman Fire Department
Firefighter Mac Nelson, Kingman, Arizona

Scottish Tacos also travel well in tinfoil to campfires, tailgate parties etc. They also warm up on manifolds of engines/generators but you gotta watch em real close. Ingredients/amounts can be adjusted to suit individual tastes but this one suited the guys at the stations I worked at just fine. I have eaten this dish from high in the Rockies of Wyoming to multiple continents and several islands in peacetime and on military missions.

GRILLED VENISON LOIN WITH ARTICHOKE GRAVY AND MASHED POTATOES

Mormon Lake is a shallow, intermittent lake located in northern Arizona. With an average depth of only 10 ft, the surface area of the lake is extremely volatile and fluctuates seasonally. When full, the lake has a surface area of about 12 square miles, making it the largest natural lake in Arizona. In particularly dry times, the lake has been known to dry up, leaving behind a remnant marsh.

1-19 oz. jar marinated artichoke hearts
½ stick margarine
½ C. flour
32 oz. chicken stock
1 C. white wine
1 lemon
¼ tsp. white pepper
2 lbs. venison loin

1. Melt butter in medium sized sauce pan.

2. Add juice from the marinated artichokes(set aside artichoke hearts to add later) and white pepper stir in flour making a rue.

3. Add the white wine, lemon juice and slowly add the chicken stock let simmer until thickened roughly chop the artichoke hearts to bite size pieces and add to the gravy.

4. Season and grill the Venison (or if you prefer you can use beef loin) to your desired temperature.

5. Slice and smother with gravy serve with mashed potatoes.

Mormon Lake Fire

MORMON LAKE FIRE DEPARTMENT
Firefighter Heather Luettjohann, Mormon Lake, Arizona

Benjamin Franklin is credited with being the "Father of the Fire Service" for his salvage operations during a fire.

Jalapeno Poppers

Located in the historic village of Mormon Lake Arizona, the Mormon Lake Fire Department was established in 1987 to provide fire prevention and emergency services to residents and visitors within the Mormon Lake Fire District. The Mormon Lake Fire Department (MLFD) is an organization comprised of 20 Volunteers protecting an area which spans over 44 square miles.

24 jalapenos
1 lb. bacon
1 lb. cream cheese
1 lb. Kraft shredded
4 cheese blends

1. Cut jalapenos in half remove seeds and veins (that is where the heat comes from.

2. Cut bacon in small strips and fry until crisp.

3. Mix softened cream cheese shredded cheese and cooked bacon.

4. Stuff mixture into jalapeno halves and bake or grill until the tops are brown.

Mormon Lake Fire Department

Firefighter Heather Luettjohann, Mormon Lake, Arizona

Fire "Plugs" or hydrants, got their name from one of the first municipal water systems. The system was made of wooden logs, and when there was a fire, a FF would drill a hole in the log to get water for the fire. When they were done, they would simply "plug" them up.

RAZORBACK JAMBALAYA

The Little Rock Fire Department is a dynamic organization comprised of twenty (20) Fire Stations covering 122.31 square-miles, protecting over 183,000 citizens. In addition, there are over 400 employees in the organization.

1 lb. smoked sausage
1 lb. shrimp
½ lb. ham or chicken cooked and cubed
1 bell pepper
2 medium onions
28 oz. can whole tomatoes
1 tsp. salt
1 tsp. ground thyme
1 tsp. black pepper
1 tsp. garlic powder
2 Tbsp. Worcestershire sauce
1 C. converted rice
Tabasco sauce to taste
¼ C. oil

1. Boil the shrimp, save the broth, peel the shrimp and set aside.

2. Dice the sausage and lightly brown in the oil. Remove sausage and set aside.

3. Sauté the diced onions and bell pepper in the sausage oil.

4. Add spices, 1 cup of the shrimp stock and rice.

5. Stir in the whole tomatoes with juice and add chicken and sausage.

6. Cook covered at a low temperature for 45 minutes adding a little water if needed.

7. Do not stir and break up the rice. Add the shrimp the last ten minutes.

Feeds 4 regular eaters or three firefighters.

"Steamers" were the first fire engines. They used steam power to build up pressure to flow water through hoses. Now days the "steamer" is the large inlet on the side of a engine to let water into the pump.

C-Cubed (Commerce California Chili)

The Los Angeles County Fire Department has a very rich and unique history, which is full of innovation, and daring accomplishments. From designing the 911 system in the 1970's to the current day USAR and Homeland Security sections, our fire department is a leader and a model to fire departments around the world. Along with our rich history, the personnel take pride in not only our organization, but also in the areas in which they serve. Part of this pride is demonstrated in the designs of their station patches, and in their station or Battalion logos. These "unofficial" patches incorporate some of the landmarks, backgrounds, and cultures of the more than 2,200 square miles we serve in the County.

Meats:
1½ lbs. Tri tip steak
1½ lbs. London broil steak
1 lb. filet mignon
1 lb. hot Italian sausage
3½ Tbsp. oil
1½ Tbsp. Gebhardt or Tampico red chili powder
1 tsp. accent meat tenderizer
1 tsp. garlic powder
1 tsp. onion powder
¾ tsp. cayenne pepper
½ tsp. salt

Batch 1:
2 (14.5 oz) cans chicken broth
1 can E1 Pato tomato sauce
1 Tbsp. onion powder
2 tsp. gar1ic powder
2 Tbsp. Gebhardt or Tampico chi1i powder
1½ tsp. white pepper
½ tsp. Cayenne pepper
½ tsp. allspice

Batch 2:
6 oz. can tomato sauce
2 Tbsp. Gebhardt or Tampico red chi1i powder
1 Tbsp. paprika
1½ Tbsp. onion powder
1 tsp. garlic powder
½ Tbsp. Tabasco sauce
1 Tbsp. beef bouillon granules (can break up if cube size)
¼ tsp. accent
½ tsp. white pepper

Batch 3:
6 oz. can tomato sauce
2 small cans diced Hatch green chiles
1 Tbsp. Tampico New Mexico chi1i powder
3½ Tbsp. Gebhardt or Tampico chi1i powder
1½ Tbsp. Tampico California chili powder
1 Tbsp. paprika
½ tsp. white pepper
½ tsp. cayenne pepper
½ tsp. tabasco sauce
½ tsp. garlic powder
1 tsp. onion powder
1 tsp. accent
1 red bell pepper, 1 green bell pepper (cut, remove seeds & chop to small pieces

Batch 4:
1 can E1 Pato tomato sauce
1 tsp. gar1ic powder
2½ tsp. brown sugar
2 tsp Cumin
1/8 tsp salt (or salt to taste)
1 lime (squeeze lime juice into pot)
2½ tsp. Worcestershire
1½ tsp. soy sauce
1 tsp. cinnamon
1 tsp. oregano
1 tsp. thyme
1 tsp. nutmeg
1 tsp. white pepper
1 tsp. garlic powder
2½ Tbsp. chipotle sauce

Meat:

1. Brown the hot sausage separately.
2. With all the meats combined immediately add 3½ tablespoons oil (sautéing) - brown meats to a somewhat semi soft knife cut & leave no pink, 1½ tablespoons Gebhardt or Tampico red chili powder, 1 teaspoon accent meat tenderizer, 1 teaspoon garlic powder, 1 teaspoon onion powder, ¾ teaspoon cayenne pepper, and ½ teaspoon salt.
3. P1ace all the above in a large stock pot / pan and simmer on very low heat. Don't use any of the liquids from browning the meat.
4. With the above, also immediately add in Batch 1.
5. Slightly stir to mix in above ingredients and stir once in a while, to keep from a hard boil & browning bottom of pot. Taste in between batches & adjust according to your preference.
6. After 90 minutes, add Batch 2.
7. Cut in half 1 Serrano & 1 Jalapeno pepper, take out the seeds & float the peppers.
8. Slightly stir only as needed to keep from a hard boil & browning bottom of pot.
9. After another ½ hour, add Batch 3.
10. Remove Serrano & Jalapeno peppers and carefully adjust need for added liquid. I use combination of a can of chicken broth, can of tomato sauce and a cup of water, to maintain plenty / proper quantity of liquid.
11. Slightly stir only as needed to keep from a hard boil & browning bottom of pot. Do not make it too thin.
12. After another ½ hour, add Batch 4.
13. To thicken as needed, slowly stir in small amounts corn starch (or follow directions on corn starch package for proper thickening. Stir until chili reaches a proper thickness. Add more corn starch if needed, to suit for your proper thickness.
14. Simmer approx. 15 - 20 minutes & slightly stir only as needed to keep from browning bottom of pot. If needed, add a bit more tomato sauce, a few pinches of cinnamon & brown sugar, to take away any type of chili bitter taste or to add more flavor.

Fire Fighter Mike Morgan, Commerce, California

Fire Departments adopted the military's ranking structure from the colonial days.

Grandma Irwin's Irish Clam Chowder

During the 1890s, fire protection for the City of Escondido was provided by a group of volunteers using a hand-drawn cart and garden hose to extinguish fires. Whenever he saw smoke along the few short blocks of Grand Avenue, the Fire Marshal would ring a small hand bell to signal the volunteers. The hose cart, made up of two buggy wheels joined by a drum, was stationed at City Hall on Valley Boulevard, standing ready on a platform serviced by a six-foot ramp.

12 slices of thick sliced bacon
8 potatoes, diced
2 leeks, chopped
16 oz. clam juice
6 C. of chicken broth
8 Tbsp. of real butter
14 oz. calms
4 Tbsp. flour
2 C. of skim milk
14 oz. of scallops
1 C. of sweet corn
20 oz. light cream

1. Start off by sautéing bacon in a large stock pot until slightly crisp, then add leeks and continue to sauté until bright green.

2. Add clam juice, chicken broth, 4 tablespoons butter and potatoes.

3. Simmer until potatoes are tender.

4. As potatoes are cooking, in a sauce pan, add remaining butter over medium heat, whisking in flour.

5. Add milk and cream until smooth, continuing to whisk until thick.

6. Combine the sauce and potatoes until well mixed.

7. Add the clams, scallops, and corn.

8. Bring to a boil and the reduce heat to simmer for 10 minutes.

9. Serve in a sourdough bread bowl (our fire house favorite.

Makes 12 servings (6 firefighters)

Fire Captain Tim Irwin, Hidden Meadows Station 3, Escondido, California

A "backdraft" is a real occurrence, not just a movie. When all the available oxygen in a room is consumed by a fire, and the heat builds up, as soon as a door or window is opened, BOOM, you get a backdraft.

Smokejumper Stew

During the 1890s, fire protection for the City of Escondido was provided by a group of volunteers using a hand-drawn cart and garden hose to extinguish fires. Whenever he saw smoke along the few short blocks of Grand Avenue, the Fire Marshal would ring a small hand bell to signal the volunteers. The hose cart, made up of two buggy wheels joined by a drum, was stationed at City Hall on Valley Boulevard, standing ready on a platform serviced by a six-foot ramp.

2½ C. pinto beans, rinsed (soak overnight)
1 can stew tomatoes
2 medium onions
3 cloves of garlic
1 small can of tomato paste
½ lb. of thick sliced bacon
1 lb. of any type of game meat (elk, venison, grouse, quail, etc.)*
1 can sweet corn
1 can of green chilies
2 fresh jalapeno, diced
1 C. of white or brown rice

*If you're a city slicker, you can add either beef stew meat or Italian sausage (sweet of hot) as a replacement

1. Start off with a large stock pot and add beans with enough water to bring to boil for approximately 2 hours (add water as needed) until tender.

2. Once beans are cooked, drain and rinse, keeping 4 cups of the original liquid and set aside.

3. Cook bacon to crisp, adding onions and garlic until tender then add to beans.

4. Bring to a boil and then add remaining ingredients.

5. Reduce heat to simmer for 2 hours.

Serves 16 or 8 firefighters.

Smokejumping was first proposed in 1934 by T.V. Pearson, the Forest Service Intermountain Regional Forester, as a means to quickly provide initial attack on forest fires.

Barbecue Ribs

Los Angeles County was one of the original counties of California, created at the time of statehood in 1850. The county's large area included parts of what is now Kern County, San Bernardino County, Riverside County and Orange County. These parts of the county's territory were split to form San Bernardino County in 1853, Kern County in 1866, and Orange County in 1889. In 1893, part of San Bernardino County became Riverside County.

3 to 4 lb. spareribs (Suggestion: boneless country-style pork ribs)
salt & pepper
1 onion, sliced
16 oz. bottle of smoky barbecue sauce (or 2 C. homemade sauce)

1. Sprinkle ribs with salt and pepper.

2. Place ribs in broiler pan under broiler for 30 minutes to brown and remove excess fat.

3. Put sliced onion in CROCK-POT. Slice ribs into serving pieces and put in CROCK-POT.

4. Pour in barbecue sauce.

5. Cover and cook on low 8 to 10 hours. (High: 4 to 5 hours)

"Truckies" are the firefighters assigned to a ladder truck and they're usually responsible for forcible entry, search & rescue, ventilation, salvage & overhaul.

German Cheese Cake

The history of the Los Angeles Fire Department is one of the most unique and inspiring in U.S. fire service history. An all-volunteer department for nearly two decades in the mid 19th Century, the department became an official agency of the City of Los Angeles on February 1, 1886. There are museums, historical archives and other resources available to anyone interested in the LAFD's past.

Filling:
1 lb. cream cheese (two bars)
3 eggs
½ C. sugar
1 tsp. of vinegar

Topping:
1 pt. sour cream
¼ C. sugar
1 tsp. of vanilla

Crust:
1 pkg. Graham crackers
¼ C. butter

Filling:
1. Bake at 375 degrees for 20 minutes – take out of oven, cool for 30 minutes.

Topping:
1. Pour on top, return to oven for 10 minutes at 375 degrees.

2. Cool, pour on any type of fruit topping – Comstock pie filling.

Crust:
1. Add melted butter to crushed crackers, spread on bottom of spring pan and store in freezer about 10 minutes to make firm.

99% of the American populations can dial 9-1-1 and summon emergency medical assistance in an emergency.

ENGINE Co 4
1900
TRUCK 4
TRUCK CO.
4

29

Lobster Marsala

In 1771, Franciscan friar Junípero Serra directed the building of the Mission San Gabriel Arcangel, the first mission in the area. On September 4, 1781, a group of forty-four settlers known as "Los Pobladores" founded the pueblo called "El Pueblo de Nuestra Señora la Reina de los Angeles", in English it is "The Town of Our Lady the Queen of Angels". The Queen of Angels is an honorific of the Virgin Mary

lobster
sweet Marsala wine
butter

1. Remove the lobster tail from the head. (I never keep the head unless the bug is 7 lbs. or more and I'm going to eat the meat out of the legs and then I only keep the legs. Personally I think the juices from the guts etc. in the head ruin the taste of the meat.)
2. Take the lobster tail(s) and spilt it/them into two halves.
3. Place the tail(s) meat side up into a pan or dish that is just big enough to hold it/them.
4. Pour sweet Marsala wine into the pan almost up to the level of the top of the meat. (Any type of red wine will work, but I have found that Sweet Marsala is the best)
5. Place thin slices of butter on top of the meat.
6. Place the pan/dish with the lobster into an oven that is on as high as it will go without being on broil.
7. The lobster will be done when the wine is boiling and the meat is a solid white. For those who would like to brown the top of the meat you can now turn the oven up to Broil and put the meat under the broiler for a couple of minutes.
8. Pull the meat from the shell and serve with drawn butter or what ever else dipping juice you like. A great dipping sauce is made by mixing in more butter in with the left over wine that the lobster was cooked in and dip the meat into that.

You will find that the lobster picks up the fruity flavor of the wine. Not only does this recipe work great with fresh tails but it works great with frozen tails too. During lobster season, I dive for lobster and I often freeze a good number of the tails that I get. This recipe works great with frozen tails as it tends to rehydrate the meat. This recipe also serves well with rice or a grain called Quinoa.

AUNTIE M'S POTATO SALAD

San Francisco (Spanish for "Saint Francis") was founded on June 29, 1776, when colonists from Spain established a fort at the Golden Gate and a mission named for St. Francis of Assisi a few miles away. The California Gold Rush of 1849 brought rapid growth, making it the largest city on the West Coast at the time. Due to the growth of its population, San Francisco became a consolidated city-county in 1856.

4-5 lbs. medium red potatoes washed and quartered
6 boiled eggs sliced thin
2 carrots grated
2 small cans sliced black olives
1 bunch cilantro washed and stemmed
2 Serrano chilies sliced thin
½ a red onion sliced thin
2 C. mayonnaise
juice of 1 lemon
1 Tbsp. red wine vinegar
6 dashes Worcestershire sauce
seasoned salt to taste
paprika

1. Boil the potatoes until done (do not over cook) drain and rinse with cold water

2. Add all ingredients in a large bowl except for eggs and paprika. Stir until well mixed.

3. Place sliced eggs on top of salad and sprinkle paprika over the top.

4. Serve with barbecue chicken or steak!

SAN FRANCISCO FIRE DEPARTMENT

Fire Chef Rafael Gordon, San Francisco, California

A San Franciscan cooking breakfast on a stove whose chimney was damaged during the 1906 quake, started the 24-hour-long "ham and eggs fire" which destroyed a 30-block area, including parts of City Hall and Market Street.

Barbary Coast T-Bone Steaks

Established in 1866, the San Francisco Fire Department is rich in tradition and history. From the Great Earthquake and Fire of 1906 to the Loma Prieta Earthquake of 1989, the Department has grown to meet the many challenges along the way. Today, the San Francisco Fire Department serves an estimated 1.5 million people, providing fire suppression and emergency medical services to the residents, visitors and workers within San Francisco's 49 square miles.

5 lbs T-Bone steaks (8 oz each)
1 pack Chili Willi Spice Blend
2 lbs. broccoli florettes (4 oz each)
3.5 lbs. Russet potatoes (6 oz each)

1. Wash, pat dry T-Bone Steaks.

2. Spice steaks, pour on top, roll over to coat thoroughly.

3. Sear 375F - 2 minutes each side.

4. Cook 350F - 4 minutes each side (medium rare).

SERVE:
Broccoli: Steamed 15 minutes (or Microwave 15 minutes)
Russet Potato: Baked 15 minutes after microwaving 15 minutes

Tadich Grill, Home of the Original Cippino

San Francisco Fire Department
Fire Chef Sig Wallen, San Francisco, California

Station 18 is in San Francisco, it has a Truck Company, ALS Engine Company and a Surf Rescue Unit. Everyone who makes the station has to be surf rescue certified and proficient in low angle rope rescue. We are responsible to respond to fire, medical emergencies, surf rescue emergencies and for people that fall off cliffs.

Borocay Beef Sandwich

Today, San Francisco is ranked 44th of the top tourist destinations in the world, and is the sixth most visited one in the United States in 2012. The city is renowned for its cool summers, fog, steep rolling hills, eclectic mix of architecture, and landmarks including the Golden Gate Bridge, cable cars, the former prison on Alcatraz Island, and its Chinatown district. It is also a primary banking and finance center.

3 lbs. flap steak (from Costco)
4 limes
olive oil
Montreal steak seasoning
worcestire sauce
1 can chipotle peppers
3-4 cloves of garlic
1 C. mayonnaise
salt and pepper to taste
6 sourdough rolls

1. Marinate the meat by rubbing olive oil, worcestire sauce and the Montréal steak seasoning, squeeze the juice of 1 lime over the top and place in a pan for an hour or more.

2. In a blender or food processor add the mayonnaise, garlic, salt and pepper, juice of 1 lime and up to 1 small can of chipotle peppers, the more you put in the hotter and spicier it will be. Blend until creamy smooth.

3. Barbecue the meat on medium heat until medium or medium rare, let sit covered for 10 minutes to rest. Carve meat thinly and against the grain and sprinkle remaining lime juice over meat.

4. Place meat in bread and enjoy!

In only 3 1/2 minutes, the heat from a house fire can reach over 1100 degrees Fahrenheit.

FRISCO RIB EYE STEAK SALAD

A stranger viewing the seal of the City of San Francisco might ascribe the Phoenix thereon to the tragic fire of 1906. But the "fire bird" had been chosen over fifty years earlier to commemorate the very birth of the City. In 1848, both dwellings and places of business were either common canvas tents or rough board shanties. The heart of San Francisco was destroyed by fire six times in a period of eighteen months. Yet, each time, following the example of its mythical symbol, the City had risen anew from its smoldering ruins.

5 lbs. rib eye steaks (8 oz each)
1 pack Barbi Q Spice Blend
2 lbs. mixed spring lettuce (4 oz each)
3.5 lbs. yams (6 oz each)
8 Tbsp. butter
2 Tbsp. balsamic vinegar

1. Wash, pat dry ribeye steaks.

2. Cut steaks lengthwise into 3" wide segments.

3. Spice steaks, pour on top, roll over to coat thoroughly.

4. Sear 375F (2 minutes each side).

5. Cook 350F (4 minutes each side (medium rare)).

6. Rest, slice with 30 degree angle into ⅓" strips.

SERVE:
Salad:
Toss salad with balsamic vinegar to coat lightly, serve steak on top.

Yams:
Baked 5 minutes after microwaving 15 minutes.

Sig Wallen likes to use Community Spice Company's Spice Blends because they are easy to use, provide big bold flavors and they are a local company to the Bay Area. The owner is a long time friend of Sig's, and donates a percentage of sales to the 911 Foundation.

Mike's Green Chili Sauce with Pork

1866 to 1869: On March 25, 1866, Volunteer Hook and Ladder Co. #1 organized, the first fire company in the Colorado Territory. The city built the first station at 1534 Lawrence St. A hand pumping draft engine was purchased in 1867 but scant water supplies and manpower limited its use.

1½-2 lbs. pork chops or shoulder meat
1 large onion, chopped
3 cloves fresh or about ¾ tsp. garlic powder
¼ tsp. cayenne powder (to taste)
24 oz. roasted Anaheim peppers, peeled and chopped
14 oz. can diced tomatoes
flour and water (to make a roux)
4 Tbsp. olive oil, or any type cooking oil

1. First, freshly roasted peppers are best in this recipe, and living in the area where I am at allows me to get these easily. You can substitute with canned green chiles but you may want to consider adding additional to the recipe to taste, because canned peppers lose a lot of their earthy taste. (If fresh peppers are available to you, you can roast them on top of the stove over an open burner on medium heat turning over frequently until the skin of the peppers is charred equally all over, beginning to peel off, and the meat of the peppers is tender. Remove (scrape off) the skin by hand or with a butter knife along with the stems, but leave the seeds that are inside the peppers. The aroma that is given off when they are roasting is unique, and you'll almost feel like you're working on North Federal Boulevard with the boys in the S.W.A.C.!)
2. Heat the oil in a large pot (about 4 quart. or just a little larger size) Add the garlic, pork that is cut into one inch (or a little smaller) cubes, and onions and fry them until the onions become transparent and the pork starts to gray and begins to brown. (If your house goes on a lot of runs, start the recipe as early in the day as you can, then transfer the mix into a crock pot and continue the recipe and let it simmer on low all day until dinnertime. For our station, we will start it around 8 a.m. and eat dinner around 6 p.m.)
3. Add the can of chopped tomatoes to the mix and add an additional one and a half tomato cans full of water. At this time also add the chopped green chilies (Anaheims). Add salt to taste and add the cayenne to raise the heat to your desired temperature (Anaheims are generally mild peppers depending on which ones you buy, so play around with it until you get it the way you like it).
4. Just before serving, add the flour and water together to make it pourable. This is to thicken the mix. This should yield about 12 servings and can go on just about anything. Scrambled or fried eggs, burritos of any kind, etc. heck, its good just by itself!
5. Top off with cheddar cheese, sour cream, lettuce, tomato and even some avocado and a few sprigs of cilantro if you' re able to get some. Enjoy!

Notes: Any and all of the ingredients can be adjusted to your particular tastes. This is just what has worked best for me, and at times a little more of one thing and less of another gets added. BE FLEXIBLE, and it's really not as hard to make as it may sound.

If you have an electric stove you can put the peppers on the top rack of the oven on broil, leave the door open and keep your eye on them. As the skin chars, turn with tongs. This only takes a few minutes. To make peeling really easy, toss them into a paper sack and let them sit and sweat a couple minutes. You'll be able to pull the skin off in big sheets.

Southwest Adams County Fire Rescue
Firefighter Mike Tavalez, Denver, Colorado

Unlike fires in the movies, the smoke from a house fire can be so thick that your house would be completely dark in 4 minutes, even with all the lights on!

Bar-B-Cups

During World War II, much wood was needed to build battleships and military transport crates. The government worried that fires could damage the forests that provided our timber. In response, the USDA Forest Service and the War Advertising Council launched a poster campaign that they hoped would help prevent forest fires.
Early posters featured Bambi, but the campaign soon switched to America's favorite toy animal—the bear. In 1944 illustrator Albert Staehle drew the first Smokey, a big-eyed, round-nosed bear in a park ranger's hat. The bear was named for "Smokey" Joe Martin, the former assistant chief of the New York City Fire Department.

1 lb. hamburger
2 cloves garlic
½ onion minced
some salt & pepper
½ green pepper or jalapeño minced as desired
mushrooms finely chopped as desired
1 Tbsp, heaping brown sugar
half C.KC Masterpiece (or your favorite) BBQ sauce
½ lb. of desired cheese
2 pkgs. of Hungry Jack biscuits (they stick the least)

1. Brown the hamburger adding all the ingredients to the mushrooms; drain & then add the brown sugar & BBQ sauce- allow simmering about 10 minutes.

2. In 2 muffin pans flatten out the biscuits against the bottom & sides to the point you can put about 1½ tablespoons [a large serving spoon] of filling in each biscuit.

3. Cover with cheese & cook as per the directions on the biscuit can- usually 400F for 10-12 minutes.

4. Figure about another 50% ingredients for each 10 pack of biscuits for bigger groups (for example add about another half pound of meat, ¼ onion & cheese, etc.) 4-7 is the typical serving; the record is 13.

In 1984 Smokey was honored with a postage stamp that pictured a baby bear hanging onto a burned tree.

Chicken Diablo

It was the great Chicago fire of Oct. 8, 1871 that "inspired" the creation of Greeley's first fire department. As you know, the Chicago fire consumed, in less than a week roughly 3.5 square miles of the city; 250 people died. Greeley's population by 1871 was ca. 1,500 and the town itself was an assortment of frame buildings which an Illinois journalist in 1870 described as "so many dry goods boxes scattered across the plains of the Almighty"----Greeley, Colorado Territory, was the perfect "tender" for a fire.

6 boneless chicken breast
½ C. hot sauce
½ C. ketchup
1 C. sour cream
¼ tsp. paprika
¼ C honey
¼ tsp. cumin
½ C. olive or vegetable oil

1. Mix hot sauce, honey and sour cream until blended.

2. Then add the paprika, cumin and ketchup and blend until it is smooth.

3. Marinate the chicken for 2 hours in refrigerator in half of the sauce.

4. Heat the oil and brown the chicken on both sides till cooked.

5. Reduce heat and cover with extra sauce until warm.

6. Serve. You can serve with mixed vegetables or a Spanish rice on the side.

Greeley Fire Department

Firefighter Christopher Benson, Greeley, Colorado

Pretzel Salad

Henry T. West, the president of the Town Board, spearheaded the establishment of Greeley's fire department, as his son's home and business had been destroyed in the Chicago fire. By Nov. 6, 1871, Greeley had adopted an ordinance authorizing the town board to appoint a fire department to recruit members, maintain equipment, conduct inspections, and enforce fire codes. The city provided funds for a fire station and equipment, but relied on members of volunteer hose companies to fight fires until professional municipal firefighters were hired in 1913. Our City Directories list these companies and their captains.

Crust:
3 C. of crushed pretzels
½ C. butter/margarine
3 Tbsp. sugar

Cream layer:
1 pkg. of cream cheese-soft
8 oz. pkg. Kool Whip
1½ C. sugar

Jell-O top:
Jell-O as frozen fruit-strawberry, raspberry, blueberry (grape), and blackberry

Crust:
1. Soften the butter & stir in sugar.

2. In a 8x12 cake pan pile the pretzels & stir the butter mixture until most all pretzels are coated. Then flatten out in the bottom of the cake pan- bake at 375F for 15 minutes.

Cream Layer:
1. Mix these ingredients with a mixer. I typically use 12 oz. of Kool Whip. You can spread this onto the pretzel crust after it has cooled (for 30 min. or so.)

Jell-O top:
1. Use the same or similar type Jell-O as frozen fruit-Strawberry, Raspberry, Blueberry (grape), and Blackberry. 12oz. bag of frozen fruit, 1- 6oz. box of gelatin.

2. Use only half of the amount of (boiling) water to dissolve Jell-O. Pour in fruit, stir slightly, & refrigerate for about 15 minutes (slightly lumpy) and spoon over creamy layer until even.

3. Refrigerate until set. (About 2 hours.)

Benjamin Franklin, Thomas Jefferson, Samuel Adams, and George Washington were in America's volunteer fire companies!

Egg in a Frame

The traditional American song "Yankee Doodle" has Norwalk-related origins. During the French and Indian War, a regiment of Norwalkers lead by Colonel Thomas Fitch. The British regiment began to mock and ridicule the rag-tag Connecticut troops, who had only chicken feathers for a uniform. Richard Shuckburgh, a British army surgeon, added words to a popular tune of the time, Lucy Locket, macaroni being the London slang at the time for a foppish dandy).

an egg
bread (any type)
drinking glass
your favorite spread (butter, margarine, or a heart-healthy equivalent of your liking - your preference)

1. On a low flame, coat the pan with your favorite spread. On a cutting board, place a slice of bread & take your drinking glass. Turn it upside down, & place it in the center of the slice of bread. Rotate the glass until the edges cut through the bread. You should be left with 2 pieces. 1 will resemble a disc, the other should look something like a picture frame.

2. Place the picture frame piece in the pan & break your egg, placing (the egg) in the hole you've created in the center of the "frame". When it's cooked to your liking (yolk broken or not), flip it over (Note: this recipe does not allow for cooking an "open-faced" egg, because the other side of the bread will not have been cooked, & the egg will not fully cook either). Continue cooking until the egg is done (the bread will be done automatically).

3. At the same time as the egg is cooking, you can place the "disc" in the pan as well, continuously flipping it over until it's browned to your liking.

Notes
1. A cookie cutter can be used instead of a drinking glass. It all depends on your preference of what shape hole you want in the middle of the bread.

2. You can place as many "frames" & discs in the pan as room allows

The first fire engine was actually a hand-operated pump for water on wheels and firefighters would often have to push and pull it to a fire.

Homemade Macaroni & Cheese

The history of the Norwalk Fire Department stretches back to the late 1770s, when British troops set fire to Norwalk. Since then, the department has had a long history of establishing organized firefighting through volunteers and eventually employees.

1 Tbsp. butter
1 Tbsp. Flour
1 C. milk
8 oz. sharp or extra sharp grated cheddar cheese
8 oz. any pasta

1. Melt the butter, and then add the flour to make"rue"(a paste-like texture).

2. Add the milk slowly to thicken the texture more.

3. Add grated cheese.

4. Pre-heat oven to 350 degrees.

5. Cook the pasta, drain, and add the cheese sauce, then bake until thickened.

6. All amounts can be multiplies depending on how much is to be made.

Norwalk Fire Department
Firefighter Martin Diamond, Norwalk, Connecticut

There are 3 common types of fire trucks. Pumper trucks are about 30 feet long and primarily rely on attaching hoses to fire hydrants for water. Tanker trucks are also about 30 feet long and can carry more than 1000 gallons of water. Ladder trucks are about 40 to 50 feet long and have a long telescopic ladder on top, called a town ladder.

Muster Team Clam Chowder

The South Glastonbury Volunteer Fire Department was established on September 28, 1927. Herbert Clark was the first Fire Chief of the South Glastonbury Volunteer Fire Department. Joseph Gordon served as the 1st Assistant Chief, with Ralph Tryon serving as 2nd Assistant Chief, Jim Killam as Treasurer and Stanley Sheffield as Secretary.

3 qt. of a pound of lean salt pork
12 C. of chopped onion
50 lbs. of chowder clams or quahogs
20 lbs. of potatoes
pinch of salt
1 tsp. of black pepper
2 C. of milk

1. Brown or fry one half to three quarters of a pound of lean salt pork cut into small cubes in a deep kettle until golden brown.
2. Pour off any excess fat that remains in the kettle. Add twelve cups of chopped onion and cook with the salt pork until tender or for about ten minutes.
3. Wash fifty pounds of chowder clams or quahogs and put the scrubbed clams in a large kettle with one or two inches of water in it. Cover the kettle and steam until all of the clams have opened up-this should take around fifteen minutes.
4. Remove the clams from the kettle but be sure save the broth. Shell the clams and chop them by hand or put them into a food processor fitted with the metal blade and process until chopped but not too finely minced.
5. Now add the clam broth to the kettle with the salt pork and the cooked onions.
6. Add twenty pounds of potatoes that have been peeled and cubed into the kettle and add more water if necessary.
7. Using the water that the potatoes were cooked in is fine and tends to give more flavor to the chowder.
8. Add a pinch of salt and a teaspoon of black pepper and simmer until the potatoes just start to soften a bit.
9. Now add the clams and simmer, covered until the potatoes are cooked. This takes around ten minutes.
10. Add two cups of milk just before you're ready to serve and add a dash of paprika for color.
11. Serve with plenty of oyster crackers.

South Glastonbury Volunteer Fire Department
Firefighter Jim Lyons, South Glastonbury, Connecticut

This recipe originated in 1992 following a successful muster and is one that we serve at our summer clam bakes. We cut down on the fat by using lean salt pork instead of the fatty kind called fat back. We cut down on the fat a little more by using milk instead of half and half or cream. Obviously this is the New England style clam chowder and not the Manhattan kind. This recipe serves two engine companies.

Chicken Florentine

In 1945, several fires occurred in the Wilmington Manor Area which included a plane crash in Biggs Field, now known as Penn Acres, and a house fire on East Grant Avenue that caused considerable damage. These fires made the citizens start thinking about a fire company of their own, which began with members of the Lions Club. In November of 1945, they announced a public meeting at the gun club on Basin Road (where the George Read School now stands) to discuss formation of a fire department. The Lions Club donated the first $10 toward the treasury for the new volunteers.

4 skinless, boneless chicken breast halves
¼ C. butter
3 tsp. minced garlic
1 Tbsp. lemon juice
1 (10.75 oz) can condensed cream of mushroom soup
1 Tbsp. Italian seasoning
½ C. half-and-half
½ C. grated Parmesan cheese
1 bag fresh spinach
8 oz fresh mushrooms, sliced
⅔ C. bacon, crumbled
2 C. shredded mozzarella cheese

1. Melt a little butter in saucepan over medium heat and sauté mushrooms and spinach. When spinach is done, remove from heat and spread on the bottom of a 9X13 dish.

2. In same saucepan, melt ¼ cup butter over medium heat. Stirring constantly, mix in the garlic, lemon juice, cream of mushroom soup, Italian seasoning, half-and-half, 1 cup of the mozzarella, and Parmesan cheese until cheese is melted.

3. Arrange uncooked chicken breasts in the dish on top of spinach and mushrooms, and cover with the sauce mixture.

4. Sprinkle with crumbled bacon. Bake 25 minutes in the 350 degrees F (200 degrees C) oven.

5. Sprinkle remaining cup of mozzarella on top and continue to bake for 10 more minutes.

6. Serve over angel hair pasta or white rice.

Wilmington Manor Fire Department
Fire Chef "Redman" Maichle, Wilmington Manor, Delaware

There are about 1.2 million firefighters in the United States. About 330,000 of them are career firefighters and about 812,000 are volunteer firefighters.

Firehouse Chicken Marinara

The City of Tallahassee's Fire Department has been fighting fires and responding to emergencies for more than 100 years. While Tallahassee developed as a city, firefighting was a community effort ranging from "bucket brigades" to hand- and horse-drawn wagons. In 1902, the first centralized fire department was established by the city, and in 1915 the first motorized truck, a LaGrange, was purchased for the sum of $8,000.

boneless skinless chicken breasts
3 x 12 oz cans of tomatoes
1 large yellow onion
4 cloves of minced garlic
fresh mozzarella
fresh asparagus
12 oz dried pasta
fresh basil, thyme, rosemary
olive oil

Marinara Sauce:

1. In a medium size pot, sauté 3 cloves of minced garlic and half of a chopped yellow onion in a tbsp of olive oil.
2. Sauté until the onion becomes translucent and then add the three cans of tomatoes.
3. Bring mixture to a light simmer, and let it cook down for about 15-20 minutes.
4. Once the sauce has reduced, remove from heat and add some fresh basil, thyme, and rosemary. Be sure to stir occasionally.
5. Season with salt and pepper to taste.

Chicken Marinara:

1. Preheat oven to 350 degrees. Heat a large sauté pan on medium heat.
2. Add about a tbsp of olive oil into the pan. Season both sides of the chicken breast, then carefully place it in the pan.
3. Allow the chicken to brown on one side, flip it, and then remove the pan from the heat. Top the chicken with thin slices of the fresh mozzarella and marinara.
4. Place the pan into the oven and allow to cook for approximately 15-20 minutes.
5. While the chicken is cooking, bring a large pot of water to a boil. While waiting on the water to boil, drizzle some olive oil onto a sheet pan and place the fresh asparagus on top.
6. Top the asparagus with the other half of the chopped onion and remaining clove of minced garlic.
7. Season with salt and pepper and place the sheet pan into the oven. The asparagus will be done around the same time as the chicken. Once water comes to a boil, cook the pasta.

Tallahassee Fire Department
Firefighter Jake Tillotson, Tallahassee, Florida

Dalmatians are known as firehouse dogs because of the important role they played in the past. Originally firefighters would drive horse-drawn fire engines and the Dalmatians would help the firefighters get to the fire faster by nipping at the heels of the horses as they ran.

FIREHOUSE RIBS

Fire prevention has been a large part of the fabric of the Tallahassee Fire Department. As early as 1843, the city enacted building ordinances that required all buildings in Tallahassee to be constructed of fireproof material. This was in response to the "Great Fire of 1843" that destroyed almost all of downtown Tallahassee, a business district of more than 90 structures. In 1912, Chief T.P. Coe was quoted in local newspapers saying that ladders should be readily available at all residences so neighbors could possibly put out a fire with a bucket of water.

2 lbs. country style ribs
Lawry's season salt
1 pkg. of bacon
1 head of cabbage
1 bottle barbecue sauce
1 head of cabbage
1 tin foil
1 bag of charcoal
1 bottle lighter fluid

1. Wash and season the country style ribs with Lawry's.

2. Season salt graciously.

3. Heat up the grill and clean the grates.

4. Place country style ribs on the Grill turning every 10 minutes until the inside of meat reaches 155 Degrees F.

5. Cook for approximately 30 minutes.

6. Brush with barbecue sauce while cooking if desired.

7. Wash and split the cabbage head in half.

8. Slice the cabbage into wedges about 1 inch thick.

9. Season each wedge of cabbage with Lawry's season salt.

10. Place 1 strip of bacon back and forth across each wedge.

11. Seal each bacon covered wedge individually in its own (sealed) foil wrapper.

12. Place all of your foil wrapped cabbage wedges on a cooking sheet in the oven for 50 minutes at 450 degrees.

TALLAHASSEE FIRE DEPARTMENT
Lieutenant Rusty Roberts, Tallahassee, Florida

To field approximately 7,000 EMS calls annually, TFD provides the largest non-hospital-based medical response force from Jacksonville to Pensacola.

AN AMAZING MEATLOAF

In the fall of 1972 after having witnessed a mobile home fire in which a citizen lost his life, resident Patrick "Pat" Scheible immediately launched a campaign to organize a fire department in Forsyth County. The result was the formation of Forsyth County Volunteer Fire Department, a non-profit organization with Jake Moore assuming the roll as our first Fire Chief and Mack Bailey as our first Assistant Chief.

1 sweet onion, finely diced (in Georgia we use Vidalia)
2 or 3 garlic cloves, minced
2 bay leaves
2 red bell peppers, or 1 red and 1 green for Christmas, cored, seeded and finely diced
2 tomatoes, halved, seeded and finely diced
¼ C. chopped fresh flat-leaf parsley
1 (12-oz) bottle ketchup
1 Tbsp.Worcestershire sauce
salt and freshly ground black pepper
extra-virgin olive oil
1 lb. ground beef (or ground venison)
1 lb. ground pork
3 eggs
leaves from 2 fresh thyme sprigs
3 slices wheat bread, crusts removed and torn into pieces
½ C. whole milk
salt and freshly ground black pepper

1. Preheat the oven to 350 F.
2. Coat a skillet with a 2-count of oil and place over medium heat. Sauté the onion, garlic (last) and bay leaves for a few minutes to a base flavor.
3. Throw in the peppers and cook them for a couple of minutes to soften. Now add the tomatoes; adding them at the point right before you turn the heat down lets them hold their shape and prevents them from disintegrating. Stir in the parsley, ketchup and Worcestershire; season with salt and pepper.
4. Simmer the relish for 10-15 minutes to pull all the flavors together. Remove it from the heat; you should have about 4 cups of relish.
5. In a large mixing bowl soak the bread pieces in the whole milk. Set aside. In a separate large mixing bowl, combine the ground beef with the ground pork and mix well. Squeeze out the milk from the bread and add the bread to mixing bowl. Add the scrambled eggs, 1 cup of the tomato relish, and thyme; season with salt and pepper. Mix well with hands. Be careful, it may still be hot.
6. Take a small baking tray and line with parchment paper. This helps with cleanup and serving. Form the meat into a loaf shape on the tray and top with another ½ cup or more of the tomato relish.
7. Bake the meatloaf for 1 to 1½ hours until the juices run clear and meat is tender—it should spring back lightly when pressed. Remove the meatloaf from the oven and let it cool a bit before slicing. Serve with the remaining tomato relish on the side.
8. You can Double this or add half again (1.5 times each ingredient) and only change is just make two loaves in a pan so they cook faster. Give you more end pieces!!)

Happy Fire house or happy family!

Fire Chef David M. Wall, Forsyth, Georgia

The first official fire department was founded in Boston, MA in 1678. But people began fighting fires with the Boston Fire Department in 1631 when the very first fire ordinance in the United States was passed. That was almost 400 years ago!

Road Kill Chili

Here, residents can swim in the ocean and cast shrimp nets in the marshes. Families can hike or bike coastal trails to wind down and disconnect. In Liberty County, tidal creeks that carried the canoes of Indians curl quietly through untouched marsh. Our tourist trade is increasing as city-weary travelers exit interstates to let tranquility lap gently at their feet. bicycling home through quiet neighborhoods.

6 lbs. Road Kill (Squirrel or Rabbit; NY Strip Steak or as a last resort -ground beef
1 large can of tomatoes or 4 large fresh tomatoes cut is cubes.
2 (15 oz. cans tomato sauce
3 C. water
2 tsp. Tabasco sauce (your choice of heat)
4 Tbsp. chili powder
1 Tbsp. oregano
2 onions, coarsely chopped
1 tsp. cumin
garlic to taste, finely chopped
2 tsp. sea salt
1 tsp. cayenne pepper
1 tsp. paprika
12 red peppers
4 yellow peppers
4 or 5 chilies
2 heaping tsp. flour
3 cans of your favorite Beer, PBR is great.

1. Sauté meat until browned.

2. Combine all ingredients except flour in a heavy pot. simmer 90 minutes.

3. Thicken chili with mixture of flour and a little water. Simmer another 30 minutes.

4. Place in refrigerator for at least a day and then reheat the next day.

Riceboro Volunteer Fire/Rescue Department
Fire Chef Dennis P. Fitzgerald, Riceboro, Georgia

94 percent of volunteer firefighters serve communities with fewer than 25,000 residents.

SPICEE
CHIKIN-
ITZ
A GUD
BURN

A favorite with the firefighters of Winder

Chicken Stew

The Winder Fire Department, under Chief Matt Whiting, responds to over 1500 emergency calls a year. Additionally, response is made to non emergency calls to assist our citizens here in beautiful Winder, Georgia. Operations consist of 2 strategically located fire stations, 3 engine companies, one ladder company, one rescue company, and one special response unit.

2 #10 cans of cooked chicken, or 4 whole chickens, cooked and deboned
1 gallon chicken broth
5 #10 cans of whole tomatoes
4 #10 cans of whole kernal corn
½ - 1 gallon whole milk
2 sticks of butter
2 onions
8 oz. ketchup
salt and pepper to taste
hot sauce, to taste

1. In a very large pot, boil chicken in salted water until done, approximately 2 hours.

2. Debone and skin chickens and discard.

3. Grind the chicken with the tomatoes and corn.

4. Mix all together and add the broth.

5. Peel the onions and add the stew mixture.

6. Add the butter and ketchup and bring to a boil.

7. Add salt, pepper and hot sauce to taste, and simmer for 3-4 hours, stirring constantly.

8. This will taste better if the rookies do all the stirring.

Note: This makes 13 gallons of stew.

City of Winder Fire Department, organized in 1908, fought fires for 15 years with horse and mule-drawn hose wagons, until the first motorized fire truck was purchased in 1923.

Firehouse Chili

During the mid-1800s, fire fighting equipment consisted primarily of buckets and portable water supplies. When Alexander Cartwright replaced W.C. Parke as Fire Chief in 1853, the department began to grow rapidly. Several new hand-drawn engine companies were added along with a hook and ladder company. In 1870, the tallest landmark in Honolulu was the bell tower of Central Fire Station, then located on Union Street. At night, a watchman would sit in the tower, ready to sound the alarm if he spotted a fire. Central Fire Station was later relocated to its present site at Beretania and Fort Streets.

4-1/2 pounds steak, trimmed of fat and sliced
1 pound ground beef
2 tablespoons olive oil
4 to 5 cloves garlic, minced
1 large onion, diced
4 stalks celery, sliced
5 14-1/2 ounce cans kidney beans
4 14-/12 ounce cans diced tomatoes
1/4 to 1/2 cup chile powder
1 heaping teaspoon pepper
1/8 cup brown sugar
1 teaspoon crushed red pepper, or more to taste
1 teaspoon garlic salt
1 tablespoon salt

1. Brown steak and ground beef in oil with garlic; drain.

2. Add remaining ingredients and bring to a boil.

Serving Suggestions:
Shredded Jack or Pepper Jack Cheese
Serve over corn bread or with tortilla chips

One of the most important pieces of rescue and fire fighting equipment in the department was purchased in 1995, a NOTAR Helicopter.

Baked Tofu Loaf

The Honolulu Fire Department, popularly known as the HFD, is the principal firefighting agency of the City & County of Honolulu under the jurisdiction of the Mayor of Honolulu. Founded on December 27, 1850 by Kamehameha III and Alexander Cartwright, the Honolulu Fire Department serves and protects the entire island of O'ahu, covering over 600 square miles (1,600 km²) of territory, home to more than 880,000 residents and over 4 million annual visitors.

(2) 21-oz containers firm tofu, mashed and squeezed to remove water
1 bunch green onions, finely chopped
2 carrots, grated
2 cans water chestnuts, chopped
2 ounces dried shiitake mushrooms, soaked and finely chopped
6 eggs, beaten
1-1/2 cups mayonnaise (Best Foods preferred)
Salt and pepper to taste
2 cans cream of mushroom soup

1. Preheat oven to 350 degrees.

2. Combine all ingredients except soup. Mix well.

3. Pour into a 9-by-13-inch baking pan and bake 80 minutes or until golden brown on top.

4. Warm soup on stove (do not dilute). Spread over top of the loaf.

5. Slice to serve.

Honolulu Fire Department
Firefighter Geoff Shon, Honolulu, Hawaii

In 1959, Hawaii became the fiftieth state of the United States. This meant that the Honolulu Fire Department had served the people of Hawaii under a monarchy, a provisional government, a republic, a territory, and finally, a state of the Union.

Greg Ramey's Green Pork Chili

Boise's fire department originally got its start when a group of 28 men signed on as the first volunteer firefighters. On April 11, 1876, a blacksmith shop owned by George Washington Stilz on Main Street was purchased as the first firehouse. The same year, a locally made hook and ladder started service as Ada Hook and Ladder Co. No. 1. In 1895, the fire station that was located at Fort Street and Eighth Street was moved to Tenth Street and Resseguie Street. This year also marked the first use of horses for the Boise City volunteers.

1 lb. ground pork
1 onion chopped
2 cloves of minced garlic
¼ C. flour
1 can each great northern beans, garbanzos and navy beans
7 oz can mild green chilies diced
12 oz bottle mild green taco sauce
4 oz can diced jalapenos (adjust according to desired heat)
2½ - 3 C. chicken broth
2 tsp. cumin
1½ Tbsp. brown sugar
salt & pepper to taste
1 level Tbsp. creamy peanut butter

1. Brown meat and sauté onion & garlic with meat.

2. Add flour to meat mixture and brown slightly, add beans(with juice).

3. Blend 2 cups of chicken broth, chilies, jalapenos and taco sauce in a blender & then add to the pot.

4. Then add all of remaining ingredients to pot. (Reserve ½ - 1 cup chicken broth to add as needed to thin mixture to desired consistency).

Serving Suggestions:
Add sour cream
Chopped cilantro
Shredded Jack or Pepper Jack Cheese
Serve over corn bread or with tortilla chips

On May 25, 1902, all of the Boise volunteers resigned, bringing to an end the volunteer status of the Boise firefighters, but signalling the beginning of the paid professionals, now known as the Boise Fire Department.

Spaghetti Salad

The first fire department was organized in the Brewery Saloon in 1885 after a fire wiped out nearly all the frame shacks along Eagle Rock Street. Ed Winn was first volunteer Chief with 20 volunteers. The first fire station was on Eagle Rock Street. That year a New Years Eve dance was held and raised $150.00 to be used to purchase a hand hose cart with 300 feet of hose. The railroad company agreed to install three (3) hose plugs to supply water for fire protection.

12 oz spaghetti
1 C. italian dressing
¼ C. red wine vinegar
¼ C. Johnnie's salad Elegance (may substitute Salad Supreme)
1 small can sliced black olives
1 medium green pepper-diced
½ medium red onion-diced
1 can (14 oz. unmarinated) artichoke hearts-diced
2 C. diced fresh tomatoes (cherry if available)
pepper and garlic salt to taste

1. Boiled spaghetti al dente, drain and blanch.

2. Add the remaining ingredients.

3. Season with pepper and garlic salt to taste.

4. Refrigerate and marinate overnight if possible.

5. Serve chilled. Add a sprinkle of fresh grated parmesan when you serve it.

Idaho Falls Fire Department
Captain Jim Freeman, Idaho Falls, Idaho

Between 1907 and 1909 the first two people on the department received pay. They were Julius Marker and Lew Tolley who were the two "drivers" of horse drawn equipment.

Garlic Basil and Honey Chicken Breast Fold Ups with Rice

The 1890's era was coming to an end and there was progress to be made. Until the community water supply was installed, no plans for a fire department were in the future. Fires were being fought by residents who dipped buckets in back yard wells. According to certain "Old Timers" fires were also fought by workers at the Illinois Iron & Bolt Company. They had a bucket brigade and would signal that a fire was in progress with a large gong located at the Illinois Iron & Bolt Company.

3 lbs. rice
chicken breast halved into six breast
3 garlic bulbs diced
1 C. honey
1 Tbsp. fresh or dried basil
2 small summer squash or zucchini
1 small onion
1 bag baby carrots
1 box rice
6 sheets aluminium foil

1. Cut chicken breast into halves. This should render about 6 halves.

2. Coat each breast with the honey. Then sprinkle on garlic and basil.

3. Cut up the squash/zucchini and onion.

4. Pre cook baby carrots until about ¾ of the way done.

5. Pre heat oven to 375 F. Lay out your sheets of foil.

6. Place breast in the middle and then surround the chicken with the veggies.

7. Fold up the foil, but make sure the top is not touching the chicken.

8. Fold up the sides. Do the same for all the rest of the pieces.

9. Place six breast on a cookie sheet and place in oven for 25-30 minutes.

10. While chicken is cooking prepare the rice. When ready place Rice on plate, place chicken breast on! rice and surround the chicken and rice with the fresh veggies.

Carpentersville Fire Department
Fire Chef Frank Ricci, Carpentersville, Illinois

On February 23, 1915 the first meeting of the Carpentersville Fire Department gathered at the Village Hall. The Village Hall was on the northwest corner of Main Street and Grove Street.

CALICO BAKED BEANS

The Elgin Fire Department was established in 1867 as an all volunteer department. The first company was the Elgin Hook & Ladder Company and held their first meeting in the city's courthouse on the evening of September 16, 1867. The Elgin Hook & Ladder Company consisted of a ladder wagon and several volunteer firefighters and operated out of the city's first fire house, a small, wood-frame building on Spring Street.

1 lb. of ground beef, browned
1 lb. of bacon, cut-up, browned and drain grease
1 large chopped onion
1 can (15 oz) butter beans
1 can (15 oz) kidney beans
1 can (15 oz) Great Northern beans
1 can (31 oz) pork and beans
½ C. brown sugar
½ C. of regular ketchup
½ Tbsp. dry mustard
2 Tbsp. molasses

1. Combine ingredients into large covered casserole dish.

2. Bake covered in oven at 325° for 1 hour.

3. Remove cover and bake ½ hour more.

ELGIN FIRE DEPARTMENT
Fire Chef John Wangles, Elgin, Illinois

The Elgin Fire Department has 133 sworn firefighters who provide a full range of coverage to 38.8 square miles and service a population of over 108,000 people.

Taco Salad

The coverage area includes the historic downtown area, two major medical facilities, two college campuses, multiple industrial and office parks, Illinois largest gaming facility, a growing commercial segment, as well as single and multi-family homes. Services are also provided along main transportation routes that include sections of Interstate 90, routes 20, 31 and 25, the Randall Road corridor, the Northwest METRA rail line, and two major freight lines.

1 head lettuce
1-2 bunches scallions
1-2 lbs. ground beef*
1 can of black olives (drained)
1 big bag tortilla chips
1 bottle Seven Seas Green Goddess or Western salad dressing
1 large pkg. shredded cheddar cheese

*enough packages of taco mix for ground beef

1. Brown ground beef. Drain excessive grease and add seasoning and allow to simmer.

2. Cut up head lettuce, scallions and olives into small pieces and mix together in a large pan or bowl.

3. Once all vegetables are mixed together, pour the ground beef into the salad mixture and mix well. Pour the bottle of salad dressing and mix well.

4. Crush the tortilla chips and pour into mixture.

5. Mix with other ingredients.

6. Add full package of shredded cheddar cheese and serve.

The Elgin Fire Department had a fantastic Fill the Boot event in 2012. Firefighters volunteered at multiple locations four Friday's in a row, collecting funds to help support the Muscular Dystrophy Association.

CHICKEN FRANCAIS

On October 10, 1895, a group of concerned citizens met at the Proviso Land Association office to discuss fire protection. The discussion was followed by a presentation to the Village of Maywood Board, and the result was the founding of the Maywood Fire Department. William Dickey was appointed Chief and three hand-drawn hose-carts were purchased to help protect the community. The hose-carts were stationed in sheds at three locations: 5th Avenue and Madison, 5th Avenue and St. Charles Road, and 19th Avenue and St. Charles Road.

2 lbs. boneless skinless chicken breasts
4 lemons
2 C. heavy whipping cream
1 C. dry white wine
½ lb. butter
3 C. flower
4 eggs
olive oil salt and pepper fresh parsley (flat)

1. Rinse Chicken Breasts and allow to dry. Lay breast flat and slice thin medalians from the top(normally 3-4 per breast ¼" thick. Scramble the eggs in a bowl and season to taste with salt and pepper.

2. Dip the medalians into the egg then coat with flour

3. Pre-heat oven to 300 degrees.

4. Fry the medalians on med-hi heat in butter and olive oil for approx 60 seconds per side to be golden brown(add butter as needed.

5. Place all cooked medalians into a large cooking pan 9" x 14" and place into oven to finish cooking through.

6. Take frying pan that was just used and add whipping cream and juice from three lemons and 2 tablespoons of butter to remains from the chicken.

7. Bring to a boil, add white wine allow to boil for 2 additional minutes.

8. Pour sauce over chicken medalians.

9. Place back in oven for 10 minutes.

10. Garnish with thin slices of remaining lemon and chopped fresh parsley.

MAYWOOD FIRE DEPARTMET
Firefighter Tony Morrone, Maywood, Illinois

In the early days of the Village, when a report of a fire came to the Department's attention, a loud whistle was blown at the pumping station, which sounded the alarm. Volunteers would run to the hose sheds where the Chief would give them information

Firehouse Deep Dish Pizza

Although the loss of property by fire was always a threat, the taxpayers saw no need for the expense of having a fire department that needed equipment and an ample water supply to save buildings. A fire in July 1874, forced citizens to think more seriously about funding a fire department. On July 6, about five minutes to midnight, the church bells once again sounded the alarm for a fire. People rushed about to help with the fire at the home of George Palmer.

Crust:
½ oz pkg. dry yeast
6 C. all-purpose flour
2 Tbsp. sugar
2 tsp. kosher salt
4 Tbsp. nonfat dry milk powder
2 Tbsp. butter (at room temperature)
2½ C. warm water

Sauce:
6 Tbsp. olive oil
1 small can tomato paste
2 medium onion, finely chopped
2 cans crushed tomatoes
4 tsp. dried basil, (fresh if preferred)
2 tsp. dried oregano, crumbled
garlic salt (to taste)
salt & fresh ground pepper (to taste)
6 large large garlic cloves, minced
1 large sweet red pepper (half finely chopped, half slivers)
1 large sweet yellow pepper (sliced into slivers)
4 Tbsp. sugar

Toppings:
1 lb. mozzarella
1 pkg. Muenster cheese slices (16 oz)
1 lb. mixed cheddar/jack cheeses
1 lb. chopped ground cooked sweet or hot Italian sausage
1 pkg. pepperoni (16 oz)
basil
oregano
sliced pepper slivers left over from sauce

Crust:
1. Take all dry ingredients and put them into a mixer with a dough hook.
2. Make sure water is warm (105-110 degrees) and add water and butter to mixture. Mix well on med to med-high speed until dough is smooth and elastic.
3. Once dough is well mixed and kneaded (about 5-8 min) remove from bowl and turn onto lightly floured surface. (preferably a wooden board) Cover with a damp towel and let rise in a warm place for approx 45 minutes.
4. While the dough is rising you can prepare the sauce. (see below)
5. Once dough has risen, you can separate approximately ¼ of it and use it for breadsticks. (just roll em out by hand and bake them with the pizza)
6. Take the remainder of the pizza dough and place it in a large, deep dish pizza pan (18") or a 11" X 13" baking dish. Make sure pan is well greased with olive oil. (you need a deep dish pan otherwise this monster will try to climb out of the pan)
7. Allow to rise for a bit longer (15 min).
8. Slowly and gently work dough from the middle to the outside edges of the pan. Pressing firmly around the edges to keep dough from pulling back to the middle.
9. Generously brush with olive oil before layering toppings.

Sauce:
1. Warm olive oil in a sauce pan on medium high heat.
2. Place chopped onion basil and oregano into pan and cook until slightly softened (about 5 minutes).
3. Empty tomato paste into pan with onion and mix well while still cooking (another 2 minutes).
4. Add chopped garlic and continue to cook (2 more minutes). smelling good now!!!
5. Now empty about 1¾ cans of crushed tomatoes along with the chopped sweet red pepper. (You may like a thinner sauce so feel free to add all the crushed tomatoes.
6. Bring mixture to boil and then reduce heat to simmer. At this point I usually add the sugar plus a little bit more basil and oregano. But that is just personal taste. Your mileage may vary.
7. Cook while the dough is rising until it reaches the desired thickness. Then remove from heat to cool.

Cheese and Toppings:
1. Start by layering the sliced Muenster cheese evenly over the entire pizza. Have as few gaps as possible. Then spread a light layer of mozzarella cheese mixed with some basil and oregano.
2. Spread the crumbled sausage over the entire pizza.
3. More mozzarella.
4. Now layer pepperonis over entire pizza.

Evansville Fire Department
Fire Chief Adam Bigge, Evansville, Indiana

As soon as the alarm was sounded, residents would call the local telephone office to find out where the fire was located. Then they would get in their cars and try to beat the fire trucks to the scene of the fire. Those without cars ran or jumped onto the running boards of passing automobiles.

Shrimp Arrabbiata

The Fishers Fire Department began the transition to a full-career department in 1989. In that time, we have progressed from an all-volunteer department to having more than 120 career fire fighters/paramedics. We provide service to more than 50-plus square miles of single and multiple-family dwellings as well as clean industrial and high-end technology parks. The Fishers Fire Department has the distinction of being the one of the first nationally accredited Fire Departments in the state of Indiana.

1 lb. angel hair pasta or 1 fresh spaghetti squash
2 lbs. raw, peeled and deveined shrimp
3 Tbsp. lemon juice
5 cloves garlic, pressed
2 Tbsp. capers (optional)
4 Tbsp. olive oil
5 Tbsp. butter
1 tsp. crushed red pepper
8 oz Kalamata olives pitted & quartered or sliced
2 (14 oz) cans diced tomatoes with basil (drain liquid from one can only)
8 oz feta cheese
8 oz fresh parmesan cheese, grated

1. In bowl or plastic zip lock bag, add lemon juice shrimp, capers and garlic. Let stand, refrigerated, for approximately 2 hours.

2. Prepare pasta or spaghetti squash. For squash, slice into two halves, long ways. Remove and discard seeds. Place on baking sheet, cut side down, and bake for approximately 1 hour on 350 oven, or until soft to touch. Using a fork, shave the squash from the rind and place in a large bowl or serving platter with sides.

3.In large skillet, add olive oil, butter and red pepper. Just when butter foams, add olives and tomatoes. Quickly bring to a boil and add shrimp. Cook until tender - do not over cook. Remove from heat and add half the feta and parmesan cheeses. Pour over drained pasta or spaghetti squash. Top with remaining cheeses and enjoy!

Fishers Fire Department
Fire Chef Dave Bobo, Fishers, Indiana

A group of concerned residences that comprised mostly of farmers founded the Fishers Community Volunteer Fire Department on March 14, 1955, after a tragic fire in a trailer that took the lives of 2 children.

Scrummy Meat Loaf

The first European thought to have visited the Matamata area was the trader Phillip Tapsell in about 1830.[3] In 1833 the Reverend Alfred Nesbit Brown visited the area and in 1835 opened a mission near Matamata Pa, but this closed the following year when inter-tribal warfare broke out.[3] In 1865 Josiah Firth negotiated with Ng□ti Hau□ leader Wiremu Tamihana and leased a large area of land, including the future site of the town which he named after the p□.

1 lb. sausage meat
1 lb. mince meat
1 C. bread crumbs
1 Tbsp. chopped parsley
1 egg
½ C. milk
2 onions
2 tsp. curry salt and pepper

Sauce Mixture:
½ C. water
½ C. tomato sauce
¼ C. worcester sauce
2 Tbsp. vinegar
½ C. brown sugar
25 g. butter
2 Tbsp. lemon juice
1 tsp. instant coffee

1. Mix ingredients together and put into dish and bake for 30 minutes, in moderate oven.

2. Pour off surplus liquid and cover with half of the sauce mixture.

3. Bake a further 45 minutes.

Sauce Mixture:
1. Bring to the boil and let simmer for 5 minutes.

2. Pour ½ over meat and retain rest to serve with meal.

Firefighter Michelle Hall, New Zealand

Volunteers are not paid because they are worthless, but because they are priceless.

Mammoth Heart Stopper Merange Cake

In 1854, the first New Zealand volunteer fire brigade was formed in Auckland. With nothing but buckets to start with, the brigade soon upgraded to manual pumps after these were provided by the Auckland City Council and insurance companies.

12 large eggs (secret... these eggs must never have been refrigerated)
1 lb. white sugar
2 Tbsp. malt vinegar
1 medium can of peach slices
½ pt. of cream

sheet of baking paper
large oven tray
very large stainless steel mixing bowl
electric egg beater (beating will last 17 minutes)

1. Separate egg whites from yolks into mixing bowl making sure NO yolk at all remains in whites [Albumen] as cake will NOT hold up.
2. Beat the egg whites for 8 minutes until the whites peak and bowl can be held upside down with whites remaining firmly onto bowl.
3. Continue beating while slowly adding sugar for a further 9 minutes.
4. This mixture will be considerable and very stiff, the bowl will be able to be inverted with no fall-out.
5. Carefully fold in the malt vinegar with minimum stirring.
6. Spatula out entire mixture onto centre of baking paper on large oven tray, smooth side to roundness leaving Pavlova 8 to 10 inches high
7. Smooth top slightly concave (whipped cream will be added in here after cooking).
8. Have oven at 100 C or 300 F slide in Merange Cake to centre of oven.
9. Cook for 1 hour. Never open oven door. Turn off oven, leave door closed and let cake sit in cooling oven for 1 hour more.
10. Remove Mammoth Cake and allow to cool for a further hour.
11. While cooling whip ½ pint of cream to just over soft whipped cream status.
12. When cooled you can crunch curved top slightly down leaving round sides as high as possible, spatula in whipped cream and arrange peach slices around whipped cream in complete circle.
13. The cake should be 6 to 10 inches high, 12 to 15 inches across depending on your cooking technique. While cake is cooking tiptoe out of the room allow no one to enter. (If seals on oven door are not good, the cake may slump; bumping oven door shut will also effect cake height.)
14. Cut cake in HUGE slices and allow anyone eating it to decide how much they will eat... it is a show stopper!

New Zealand Volunteer Fire Brigade
Fire Chef Murray Jamieson, Christchurch, New Zealand

"The fire didn't have organizational problems." (from his account of Mann Gulch in Young Men and Fire)
Norman MacLean

MATARIKI PORK AND VEGETABLES

In Wellington, the Town Protection Act was passed by the 'Wellington Provincial Council'. This act required each householder to keep two buckets of water and supply them when fire broke out. A manual fire engine was bought by the council and put at the Thorndon Police Station – under the control of the police.

3 Tbsp. olive oil
4 cloves garlic, finely chopped
1 Tbsp. toasted fennel seeds ground coarsely
2 Tbsp. toasted coriander seeds ground coarsely
zest of 1 lemon
1 dried red chilli, thinly sliced
1 bay leaf
1 onion, finely chopped
1 small carrot, finely chopped
2 kg. shoulder of pork roast, boned, skin scored to the flesh in parallel lines about 2 cm. apart
200 mL. white wine
500 mL. beef stock
2 purple kumara, peeled and diced 4 cm.
8 small potatoes, peeled
6 silver beet leaves, sliced thickly, boiled for about 5 minutes, cooled under cold water, squeezed dry
2 Tbsp. corn flour, dissolved in 3 Tbsp. cold water
salt and freshly ground black pepper

1. Preheat the oven to 190° C. In an ovenproof dish, heat the oil over moderate heat and add the garlic, fennel and coriander seeds, the zest, chilli, bay leaf, onion and carrot. Fry gently for about 10 minutes until the onion is soft.

2. Add the piece of pork. Turn so that it colours all over. Add the wine and let it bubble for a few minutes then add the stock. Bring it to the boil. The liquid should come about halfway up the pork. Cover and place in the oven for one and half hours.

3. Uncover, skim the fat from the liquid, add the kumara and potatoes, recover and place in the oven for a further hour.

4. Remove from the oven. Remove the pork and vegetables to a serving dish.

5. Place the casserole on the heat and bring to the boil.

6. Add the silver beet leaves, bring back to the boil, and then stir in just enough of the corn flour mixture to thicken lightly.

7. Taste and season. The pork will be very tender so slice it up and serve with vegetables and sauce.

"Human error is a consequence not a cause. Errors are shaped by upstream workplace and organizational factors….. Only by understanding the context of the error can we hope to limit its reoccurrence".
James Reason

BILLBOARD BOB'S BEER CAN CHICKEN

The Vancouver Fire Department was established in 1886, and within sixteen days of its existence the City of Vancouver burned to the ground. By 1911 the department was one of the region's best and by 1917 it was completely motorized (no more horse drawn equipment). Since 1893, 48 Vancouver firefighters have died in the line of duty.

1 whole roasting chicken (2kg/about 4 lbs.)
2 cloves fresh garlic, minced
30 ml. seasoning salt (Montreal steak spice, Lawreys, Hy's, etc.)
2 Tbsp. fresh ground black pepper
1 can your favorite beer

1. Preheat barbeque to 190° C/375°F
2. Open the beer and pour half for yourself. Reserve the half can for the chicken.
3. Rinse the chicken and allow to drip dry. Place the chicken in a small roasting pan and rub the inside with the garlic and half the seasoning spice. Season well with cracked pepper.
4. Rub the outside of the bird with more of the seasoning spice and pepper.
5. Carefully insert the beer can into cavity between the legs. The chicken should stand up, balanced on the beer can and its legs. Place the pan on the grill of the barbeque and close the lid.
6. Cook for about 1 hour. You can add 250 ml/1 cup wood chips soaked in water and wrapped in tinfoil (punch holes in the top to allow smoke to escape) for an extra smoky flavor.
7. When chicken is browned and crispy, remove from grill and allow to rest for 5 minutes. Carve the chicken and serve the warm beer and juices left in the can as a light sauce.

Canada has about 10% of the world's forests. Each year over the last 25 years, about 8300 forest fires have occurred.

Randy's # 14 Ball Thai Noodle Salad

VFRS has 2 aluminum fire boats, Fireboat two and five. They were designed by naval engineering firm Robert Allan Ltd. They are two of 3 shared boats by municipalities within the Port of Vancouver area. Port Moody and Burnaby share the other boat. North Vancouver City and District Fire Departments used to each have a boat until 2011 when they cut the service.

Dressing
5 Tbsp. fresh ginger, minced
2 Tbsp. fresh cilantro, chopped
1 Tbsp. brown sugar
½ C. soy sauce
2 Tbsp. sesame oil or to taste
1 Tbsp. crushed chili peppers or to taste
1 Tbsp. Dijon Mustard
¾ C. mayonnaise
1 pkg. or noodles, cooked and cooled (or egg noodles)
1 large red pepper, roasted, skinned and sliced
1 large yellow pepper, skinned and sliced
1 large orange pepper, skinned and sliced
4 green onions chopped
2 C. snow peas, trimmed, cooked lightly and chilled
½ C. slivered almonds for garnish

1. In a mixing bowl, combine the ginger, cilantro, sugar, soy, sesame oil, chili peppers, mustard and mayonnaise.

2. Mix well until smooth.

3. In a salad bowl, combine the cooked noodles, roasted peppers, green onion and snow peas.

4. Toss well and add enough dressing to evenly coat all the noodles.

5. Top with slivered almonds.

Fire suppression costs over the last decade in Canada have ranged from about $500 million to $1 billion a year.

Grandma's Little Applecakes with Vanilla Ice Cream

The Feuerwehr (German for Fire defense) is a number of German fire departments. The responsible body for operating and equipping fire departments are the German communities ("Gemeinden") and cities ("Städte"). By law, they are required to operate a fire-fighting force. In cities, this is usually performed by the Fire Prevention Bureau, one of the higher-ranking authorities. Most of Germany's 1,383,730 fire fighters are members of voluntary fire brigades.

20 g yeast
250 g milk
150 g flour
50 g hazelnuts
50 g hazelnuts
4 pcs. apples
2 large spoons sugar
1 bit salt
4 pcs. eggs bit of butter for baking
2 large spoons sugar
1 bit of cinnamon of top for later
1 Vanilla ice cream (Haagen Daaz is our favorite)

1. Put the yeast into the lukewarm milk and dissolve.

2. Give the flour into a bowl put a little hole in the middle of it and fill in the yeast-milk-mix and stir.

3. Peel the apples, take the pips out and cut into small cubes.

4. Put the apples and the rest of the ingredients into the flour and mix everything.

5. Cover the bowl and put it in some dry and warm place for 20 - 30 minutes allow the paste to inflate.

6. When the paste has grown take a spoon and work out small cakes of the paste.

7. Heat up a pan, put in the butter and bake the small cakes.

8. Top the cakes with sugar, cinnamon and ice cream and serve the stuff while it's still hot.

Wurttemburg Fire Department
Firefighter Tobias Setzer, Wurttemburg, Germany (International)

All fires or emergencies requiring assistance from the fire service can be reported using the toll-free number "112" in Germany.

Red-Hot Chili Korma (Curry)

Prior to 1940, the public fire service in the UK was, by today's standards, disjointed, inadequately trained, and under funded. Every small town had its own fire brigade with its own Firemaster, fire station - or stations in the cities and larger towns - and its own way of doing things. However, the onset of the Second World War changed political thinking in the area of 'home defence' and the existing diverse fire services were nationalised to become the National Fire Service (NFS).

675g lean leg of lamb, off the bone
3 Tbsp. ghee or oil
2-4 Cloves of garlic
8oz (225g) red onion, chopped garam masala
salt to taste

Marinade:
1 Tbsp. tomato puree
1 Tbsp. paprika
4fl oz. (100ml) red wine (optional)
2fl oz. (50ml) bottled beetroot vinegar
1 beetroot, ping-pong ball size, sliced
20 fresh deep-red chilies or dried kashmiri chillies, seeded and coarsely cut
1 red pepper, seeded and coarsely cut

1. Cut the meat into cubes about 1½ inches (4 cm).

2. Put into a blender the tomato puree, paprika, red wine and beetroot vinegar. Mulch into a loose paste.

3. In a large non-metallic bowl combine the paste with the rest of the marinade ingredients.

4. Add the meat and coat well.

5. Cover and refrigerate for 12-24 hours. Heat the gee or oil stir-fry the garlic for one minute, then the onion for about 10 minutes Using a 4-5 pint casserole with lid, combine the fried ingredients with the meat and marinade and place into an oven preheated to 375F, 190C, Gas 5 After 20 minutes, inspect and stir adding water if it's becoming too dry.

6. Repeat 20 minutes later, adding the garam masala and salt to taste.

7. Cook for a further 20 minutes or cooked to your liking. If you prefer to cool this one down reduce the number of chilies used.

Strathclyde Fire & Rescue
Firefighter Ian Allan, Johnstone, Strathclyde, UK (International)

Strathclyde Fire & Rescue is the second largest emergency service in Europe. It has 111 strategically sited fire stations across the Strathclyde region of Scotland

CARIBBEAN RICE PORRIDGE

Under the Police Act 1961, chapter 142 of the Laws of the Virgin Islands and associated Subsidiary Legislation, the Royal Virgin Islands Police Force was made responsible for Firefighting in the Territory.

4 C. water
½ tsp. of salt
1 C. rice
1 can evaporated milk (12 oz)
1 Tbsp. vanilla essence
¼ tsp. ground nutmeg/cinnamon
4 Tbsp. of sugar
sliced fruit

1. Add salt to water and bring to a boil.

2. Add rice, occasionally stirring slowly.

3. When rice is almost cooked, add milk, essence and spice.

4. Stir slowly until fully blended.

5. Add sugar. Continue stirring slowly.

6. Let cook until rice has absorbed ingredients, and the individual grains of rice can be clearly seen.

7. Let cool; serve alone or with fruit.

VIRGIN ISLANDS FIRE AND RESCUE SERVICE DEPARTMENT
Sub Officer Elvin E. Forbes, Virgin Islands

It has been said that Caribbean Rice Porridge is a dish of Spanish origins, brought to the Western Hemisphere hundreds of years ago. Although it was first created to be a dessert, many young West Indians have enjoyed it as a meal.

Veggie Pasta

This is one of the marvelous dishes created by Philip Glasgow, the Virgin Islands Fire and Rescue Service's very own budding chef and Fire Officer. Philip is better known for the delicious Indian Rotis he makes. However, he often needs to prepare a quick meal when on the job at the Fire Headquarters located in Road Town, Tortola. Both preparing the ingredients and cooking them can have this Veggie Pasta dish done in about half an hour. It is a quick, healthy and rather fancy looking way to enjoy a meal between calls. A pot, a few utensils and ingredients and a little time is all that is needed to satisfy a hungry Firefighter with Chef Glasgow's Veggie Pasta!

1 lb spaghetti/pasta
2 Tbsp. olive oil
1 Tbsp. salt
1 Tbsp. pepper
1 C. chopped pumpkin
1 C. chopped carrots
10 oz chopped spinach
1 chopped onion
1 clove of garlic
1 C. mushrooms
7 oz coconut milk
½ C. parmesan cheese
1 large tomato (diced)
garlic bread

1. Boil 2½ - 3 cups water in medium sized pot.

2. Add salt, pepper, pasta, onions, olive oil, carrots, spinach, garlic and mushrooms.

3. Strain away water when pasta is almost cooked.

4. Add coconut milk, tomato and cheese.

5. On a low fire, stir ingredients to prevent sticking to pan.

6. Stir for 5 minutes, and then turn off stove.

7. Let cool. Serve with Garlic Bread.

Virgin Islands Fire and Rescue Service Department
Fire Officer Philip Glasgow, Virgin Islands

"When your attitude is sick, stay home, or at least stay away from me. I don't want your emotional vomit on me."
Chief Rich Gasaway

Italian Potato Soup

On February 4, 1869 the city of Cedar Rapids began organizing its first volunteer fire department. On March 1, 1869 the city's first Silsby Steamer Engine and 500 feet of hose arrived at a cost of $6,000.00. By March 11, 1869 the fire department became fully organized with J.J. Snouffer as the foreman.

1 lb. ground Italian sausage
1 lb. ground hamburger
1 tsp. crushed red peppers
1 large diced white onion
4 Tbsp. bacon pieces
2 cans (14.5 oz) of chicken broth
1 C. heavy whipping cream
1½ lb. sliced Russet potatoes, or about 3 large potatoes
1 small bag of broccoli bits

1. Sauté Italian sausage and hamburger and crushed red pepper in a large pot. Drain excess fat, refrigerate while you prepare other ingredients.

2. In the same pan, sauté bacon, onions and garlic over low-medium heat for approximately 15 mins. or until the onions are soft.

3. Add chicken broth and heat until it starts to boil.

4. Add the sliced potatoes and cook until soft, about half an hour.

5. Add the heavy cream and just cook until thoroughly heated.

6. Stir in the sausage and the broccoli bits, let all heat through and serve.

Firefighters' clothing can weigh up to 60 pounds!

Mom`s Meatballs

The town of Fayette was platted by Samuel H. Robertson in January 1855. In May of the same year, construction began on College Hall (now Alexander-Dickman), the original building of Upper Iowa University. Upper Iowa is presently the fastest growing, private, co-educational liberal arts university in Iowa. UIU is only one of three universities to offer a three-year bachelor degree program in the United States and the first to do so west of the Mississippi.

2 lbs. hamburger
¾ C. milk
¼ C. barbecue sauce
2 eggs
1½ C. graham crackers (crushed)
salt&pepper (to taste).

Sauce for Top:
1 can tomato soup
¼ C. vinegar
1 C. brown sugar
1 tsp. dry mustard

1. Mix the meat, milk, barbecue sauce, eggs, graham crackers sat & pepper together.

2. Make meatballs and put in a 9x13 pan and cover with sauce.

3. Bake for 1 hour in a 350 degree oven.

Firefighter Scott Luchsinger, Fayette, Iowa

Fayette has been honored with a citation as "Tree City USA" for 16 consecutive years.

Truckie Barbecued Meatballs

Ethel Hunt's "History of Paola, Kansas" tells that the fire wagon was horse drawn and firemen all put on red shirts before going to a fire. Water was pumped from nearby wells. On 1874 a disastrous fire ripped through the St. Charles Hotel and Union Block. Considered at the time the worst fire in the City's history. A large number of business rooms were destroyed with an estimated $75,000 loss.

1 can evaporated milk
3 eggs
3 lbs. ground chuck can substitute a mix of 50/50 with ground turkey (for the low fat freaks)...no tofu or rice cakes
2 medium onions or 1 large one
1 round thing of cheap oat meal
garlic powder
1 good sized plastic bottle of ketchup
brown sugar
liquid smoke
tin foil

Balls:
1. Pre heat the oven to about 350 or 400 depending on how long you can hold them off.
2. Find a big old pan about a couple inches deep and say 12"by 18" or so and wrap it in tin foil for an easy clean up (not critical when Rookies are in the house).
3. Wash your hands good or glove up (critical if you have any Howard Hughes types around).
4. Get a big old bowl and dump in the beef, 1 finely chopped medium onion or ½ a large one, 3 eggs, around 2 cups of oat meal, ¾ can or so of evaporated milk and about a ½ teaspoons of garlic powder, in no particular order. Now don't hesitate jam your hands in and get to mix in.
5. Now get out that old ice cream scoop you save for special occasions and dig in. Make balls any size you want and pop'em in the pan. I make them Ice cream scoop size cuz it's fast and when you feeding wild animals you don't fiddle around or you'll get bit.

Sauce:
1. Now get another bowl and dump in a full bottle of ketchup that's right the whole bottle.
2. Now you should know why I said a plastic bottle of ketchup. If not get to shakin that glass bottle.
3. Now add a about ½ cup of brown sugar, ½ tsp. garlic powder, 1 coarsely chopped onion and about 1 tablespoon liquid smoke. Now mix with a fork or some fingers what ever is handy. Get it *handy*. Wa La! Truckie barbecue sauce.
4. Now spoon the mix over the top of the meatballs and slap it in the oven for about an hour. Drain the juice and serve in the pan. We usually serve this chow with big ole baked tators and green bean casserole.

Paola Fire Department
Chief Andy Martin, Paola, Kansas

One of the more interesting facts about fire is that most house fires start in the kitchen. Cooking is the leading cause of home fire injuries. Cooking fires often start from overheated grease and unattended cooking. Electric stoves are involved in more fires than gas ones.

Cream Tacos

The Wichita Fire Department was organized in 1886 when Wichita outgrew the volunteer fire companies which had protected the City since 1872. Today, the Wichita Fire Department has evolved from its original purpose of fighting fires to its current task of providing a much broader level of emergency services to the public. Over the years, this expanded role has allowed the Department to demonstrate exceptional community value.

2 lbs. ground beef
1 lb. brick chili
1 can diced Rotel tomatoe
1 lb. Velveeta cheese
1 can pinto beans and
½ pt. whipping cream

1. Brown & drain hamburger.

2. Pour into crock pot.

3. Combine other ingredients into crock pot.

4. Cook on low for 3 hrs.

5. Serve over corn chips.

6. Top with cheese, lettuce, tomatoes, onions and sour cream.

Wichita Fire Department
Firefighter Monty Ayala, Wichita, Kansas

Engine House No. 6 was the last horse drawn station in Wichita. By 1918 the horses were gone, and Wichita had become the first all-mechanized fire department in the United States and the second in the world.

NACHO POTATO SOUP

The Wichita Fire Department serves a resident population of over 382,000 people, and is comprised of 412 uniformed and 11 civilian members serving within the Department's two divisions. These members proudly serve the community by responding to fires and other emergencies from the city's 22 firehouses. In 2010, the Emergency Operations Division provided over 68,000 unit responses to over 44,000 calls for service.

1 box Au Gratin potatoes
1 can drained corn
1 can Rotel
dash of Tabasco
2 C.water
2 C. milk
2 C. Velveeta cheese - cubed

1. Combine all ingredients in a pot.

2. Simmer for 30 min.

Kansas Firefighters Museum

WICHITA FIRE DEPARTMENT
Firefighter Monty Ayala, Wichita, Kansas

Arson is the third most common cause of home fires. Arson in commercially operated buildings is the major reason for fire deaths and injuries in those types of properties.

3RD PLATOON STUFFED CHICKEN

The Danville Fire Department was established in 1876 and is one of the oldest paid fire departments in the State of Kentucky. Currently, the department is organized with two staffed fire station). Firefighters are on duty 24/7, 365 days per year at both fire stations. Our motto: You call....we respond!

4 thick boneless skinless chicken breast
cream cheese whipped 12 oz
1 pkg. Oscar Mayer bacon pieces
McCormick perfect pinch rotisserie chicken seasoning

1. Mix cream cheese and bacon in bowl; sit aside

2. Take chicken breast making a single 2" cut horizontal to make opening in breast

3. Using fingers stuff cheese/ bacon mix and stuff chicken to desired amount

4. Season chicken heavily

5. Place chicken in a glass dish, cooking at 400 degrees F

6. Season multiple times throughout cooking

7. Cook until chicken is firm and thoroughly cooked

8. Serve with sides of your choice to have a filling meal.

Chief said he wanted to see foam

Recipe was thought of standing in grocery store after clearing from a call. We are a close knit shift and cook everything together as a family every shift, always making a new challenging dish and having good competition between us.

DANVILLE FIRE DEPARTMENT
Fire fighter David Spanyer, Danville, Kentucky

Christmas Cookies

On October 30,1991 Neon fire Dept was called to a forest fire that was endangering a structure. Firefighters John Adams and John Emerson Spangler were among those who responded to the call. They had grown up together in the community of Mayking and were best friends. It was a warm windy night -- which became a major factor regarding the fire. Due to the magnitude of the incident, several departments responded to assist.

3 C. flour
1 tsp. baking powder
¼ tsp. salt
1 C. shortening
3 eggs
1 tsp. vanilla
1¼ C. sugar

1. Sift dry ingredients.

2. Add shortening and mix with a fork.

3. Add eggs and vanilla; mix.

4. Roll out thin and cut into desired shapes.

5. Bake at 375 degrees for 8 minutes.

6. Should yield about 4 dozen cookies.

7. May be painted, decorated or iced.

Volunteer fire fighters comprise 73% of fire fighters in the United States.

Corn Chip Salad

A gas pocket erupted and the men were overtaken by flames.John Adams outran the flames but turned to help Emerson after he had fallen. Firefighter Adams was flown to the U.K. Medical Center Burn Unit with burns over 90% of his body. He later died of his injuries. Firefighter Spangler died at the scene. His body was recovered the following day. They were the first members lost in the line of duty since the Department was created in 1947.

2 C. whole kernel corn
2 C. shredded Velveta cheese
1 C. mayonnaise
1 C. diced green pepper
½ C. chopped onions
1 bag chili cheese Frito's

1. Mix all ingredients except chips in bowl.

2. Chill.

3. Stir in chips just before serving

4. Reserve a few chips to sprinkle on top.

Wow, the engineer looks small from here!

Tom Haynes has been a member of the Neon Volunteer Fire Dept since 1975. He was Asst Chief for Rescue from 1976 - 94 and Captain Training 1994 - present The Neon Volunteer Fire Dept is located in the coalfields of Eastern Kentucky.

Chicken Jambalaya

In March 8, 2001, the Citizens of Gonzales overwhelmingly approved a 1/2 Sales Tax to improve Fire, Police and Sanitation Services. The Mayor and City Council immediately employed a full-time Fire Chief to begin the vision for the development of a professional fire and rescue department. Today just nine years later the promise was delivered and the following successes exist:
The Department has improved to a "Class 2" Fire Rating up from the "Class 4."

1 4 to 5 lb. hen cut into serving pieces
2 tsp. granulated garlic
3 C. long grain rice-uncooked
½ C. green onions-chopped
1 C. cooking oil
¼ C. bell peppers-chopped
3 C. onion-chopped fine
½ tsp. black pepper
3 tsp. salt (or salt to taste)
2½ tsp. Louisiana hot sauce

1. Fry chicken in cooking oil until dark brown.

2. Remove cooking oil leaving just enough to cover the bottom of pot.

3. Add onions and cook with chicken until onions are dark brown.

4. Add about ½ cup of water and then add bell peppers and green onions.

5. Let simmer for about 10 minutes.

6. Add about 5 cups of water and remaining seasoning.

7. Then bring water to a rolling boil. Add rice and stir.

8. Let rice and water boil until water thickens and turn rice over once.

9. Do not stir as this will break up the rice. Cover with tight fitting lid and let simmer on a low fire for about 10 more minutes.

10. Turn (as in fold) one more time and let set for about 10 more minutes and then your jambalaya is ready to eat.

Gonzales Fire/Rescue
Fire Chief Preston P. Landry, Gonzales, Louisiana

Another fact about fire is that smoking is the primary cause of death by fire in the U.S. The second cause of fire deaths is heating equipment.

Lobster Thermidor

The Hose 5 Fire Museum was built in 1897 and used for close to 100 years before the replacement fire station was built on the Hogan Road. Today, the former fire station is used as a fire museum at Bangor Maine.

3 live lobsters (abt. 1½ lbs. each)
9 Tbsp. butter
3 Tbsp. heavy cream
2 Tbsp. chopped shallots or onions
1 tsp. finely chopped chervil or parsley
¼ tsp. cayenne pepper
2 Tbsp. whipped cream
1½ C. white sauce
⅔ C. chopped mushrooms
white wine
½ tsp. Worcestershire sauce
½ tsp. dry mustard
1 egg yolk, slightly beaten
2 Tbsp. grated parmesan cheese

1. Take the live lobsters and kill them by putting them into a pan of boiling water. Cut completely through the shell to divide lobsters into halves, disjoint large and small claws.
2. Using a large heavy skillet with a tight fitting cover, heat 6 tablespoons of the butter and the lobster halves, meat-side down. Place large and small claws on top. Cover. Cook slowly 12 to 15 minutes or until tender.
3. Meanwhile, prepare white sauce (note that the recipe only makes 1 cup, so you will need to increase the recipe by 150%) and set aside.
4. Clean and cop mushrooms. Heat 3 tablespoons of butter in a saucepan and add the mushrooms. Add chopped shallots. Cook over medium heat until mushrooms are tender and lightly browned. Occasionally move and turn mixture with spoon. Remove from heat. Blend into ½ of the white sauce, heavy cream, 2 tablespoons of white wine, parsley, Worcestershire sauce and mixture of dry mustard, salt and cayenne pepper. Add to the mushroom-onion mixture and cook over low heat until thoroughly heated, stirring frequently.
5. When the lobster is done, starting at the tail, gently pry the lobster meat from the shells, reserving the shells. Remove the meat from the large claws. Place the shells, cavity side up, on a baking sheet at heat at 325 degrees about 7 minutes or until shells are heated through. Meanwhile, cut the lobster meat into 1-inch pieces and blend into the sauce. Remove the shells from the oven and sprinkle white wine (about ¼ teaspoon per shell) over the interior of each.
6. Fill lobster shells with the lobster mixture. Pour remaining white sauce into the top of a double boiler, stirring over low heat until well heated. Vigorously stir about 3 tablespoons sauce into egg yolk. Immediately return mixture to top of double boiler. Stirring constantly, cook over simmering water 3 to 5 minutes. Remove from heat and blend in whipped cream. Spoon over lobster mixture into the shells. Sprinkle 1 teaspoon of parmesan cheese over each of the filled shells.
7. Place on baking sheet on broiler plan with tops of food 2 to 3 inches from the burner. Broil 2 to 3 minutes or until lightly browned. Serves 6.

Retired Firefighter Dana Asdourian, Bangor, Maine

All 911 calls are recorded for accuracy as well as to log all times and dates. This is done automatically by machine for obvious reasons. All fire department radio traffic is also recorded and logged for times and date.

DRIED CHERRY CHILI

One of the most visible MFS activities is the prevention and suppression of forest fires. Some folks still remember the widespread devastation resulting from the fall 1947 fires, which brought about significant and positive change to the MFS that have continued over the ensuing decades. Through upgraded training, improved field communications and the reliance on an air fleet to knock down fires quickly, acreage lost to wildfires has been reduced to about 400 acres annually.

2 C. reduced sodium chicken broth, divided.
4 oz dried tart cherries (chopped ¾ cup) (or use dried cranberries)
1 Tbsp. olive oil
1 C. chopped onion
1 Tbsp. chopped garlic
2 tsp. chopped jalapeno
1 lb. ground turkey
1 roasted red bell pepper, cut into ¼" cubes (or substitute fresh green beans)
1 Tbsp. plus 1 tsp., chili powder
1½ tsp. ground cumin
1 tsp. ground coriander (or omit)
1 tsp. dried mustard
½ tsp. oregano
4 C. fire-roasted tomatoes (or any tomatoes)
2 (16 oz) cans black beans
¼ C. chopped cilantro

1. Add 1 C chicken broth to cherries and set aside.

2. Sauté onion in olive oil, add garlic and jalapeno, add turkey.

3. Cook until no longer pink.

4. Add remaining ingredients, bring to a boil, simmer uncovered for 5 minutes.

Firefighter Gregg Hesslein, Maine

Four out of five wildfires are started by people, nature is usually more than happy to help fan the flames. Dry weather and drought convert green vegetation into bone-dry, flammable fuel; strong winds spread fire quickly over land; and warm temperatures encourage combustion.

PIZZA MEATLOF

In 1787 Portland's first organized Fire Company was formed and was named the "Neptune". More companies followed: "Vigilant" in 1794, "Cataract", "Portland", "Extinguisher", and later the "Alert". Hand Engines came into service starting with the "Deluge" in 1827 and the "Hydraulion" and "Niagra" in 1830. And in 1827 the "Washington Hook and Ladder" became the City's first Ladder Company.

5 lbs.of hamburger meat (recommend 85/15)
1 large onion
1 lb. shredded cheddar cheese
1 box of Italian break crumbs
3 large peppers (green, orange and yellow for color)
5 cans of pizza sauce (5 to 6 oz size)
5 eggs

1. Dice the onion and peppers.

2. Add the onion/pepper mixture to the raw Hamburg and the 5 cans of pizza sauce. Add the egg and breadcrumbs and mix until well mixed.

3. Bake in a 350 degree preheated oven for 60 to 90 minutes (until the top browns).

4. Cover the top of your meatloaf with shredded cheese and return to the oven for 10 minutes.

5. Serves 10 firefighters with healthy appetites!

PORTLAND FIRE DEPARTMENT
Retired firefighter Dana Asdourian, Portland, Maine

Fire protection started in Portland back in 1768 with the appointment of seven "Fire Wards." These men had Police powers and had the authority to order citizens to help at Fire scenes.

Sundried Tomato, Basil and Bacon Chicken

We are a combination system of nearly nine-hundred career and volunteer providers operating from twelve stations across the County. The department is located between Baltimore City and the District of Columbia and provides automatic aid to our surrounding partners in Prince George's, Anne Arundel, Baltimore, Carroll, Montgomery and Frederick County.

Howard County Fire and Rescue
Firefighter Kevin Weisenborn, Glenwood, Maryland

8-10 boneless skinless chicken breasts
1 lb. hickory smoked bacon
3 large shallots (sliced)
1 C. fresh basil (chopped)
½ C. sundried tomatoes (sliced)
¾ C. butter or vegetable oil spread alternative
2 cloves of garlic (minced)
¼ C. grated parmesan cheese
salt/pepper to taste

Brine Solution:
6 C. water
¼ C. salt

Prep 1 Hour or More Before:
1. Make a brine solution by mixing a ¼ cup of salt with about 6 cups of water in a large bowl.
2. Submerge chicken in brine solution and place in the refrigerator for about an hour.

Sauce/Topping:
1. Chop precooked bacon into squares and place in a large skillet at medium/high heat.
2. When almost fully cooked add sliced shallots until golden in color then reduce heat to low. Next add the Butter, Sundried Tomatoes, Grated Parmesan Cheese, and ½ cup of the Fresh Basil stirring occasionally. Retain the other ½ cup for garnish.

Grill/Oven:
1. Preheat grill to medium/high heat.
2. Remove breasts from brine solution and rinse under cold water. Lightly coat grill with oil and add the chicken.
3. Preheat Oven to 350 degrees.
4. When Chicken Breasts are cooked thoroughly, place on a baking sheet and spoon the skillet mixture over top and bake an additional 10 minutes in oven at 350 degrees. Remove from oven and sprinkle remaining Basil on top.
5. Goes great with pasta dishes.

Firefighter Kevin Weisenborn joined the department six years ago and is now currently stationed at our newly-opened station, Glenwood Fire Station 13. Before working at this station he was known as the head chef at his previous station. He was never trained as a chef but from an early age of 12 started watching cooking shows on TV.

Special Operation 10
Roasted Rosemary Chicken

Howard County is located between Baltimore and Washington in the "heart" of Maryland and was named after John Eager Howard, a Revolutionary War hero and the fifth Governor of Maryland.

10 chicken breast
4 oz of olive oil
2 Tbsp. garlic minced
4 Tbsp. rosemary chopped
salt and pepper to taste
1 Tbsp. garlic powder
1 Tbsp. old bay

1. Wash chicken breast.

2. Combine oil, garlic, rosemary, salt, peper, garlic powder and old bay.

3. Roll the chicken in this mixture until they are evenly coated.

4. Sauté chicken breast till golden brown both sides in the pan.

5. Place chicken on a sheet pan place in oven cook till done.

Firefighter Barry Griffin, Glenwood, Maryland

FF Griffin is a professionally trained chef from the Pennsylvania Institute of Culinary Arts. He studied the profession from 1997-2003, before joining the Department in 2003. In the special operations team, FF Griffin cooks most of the time and has been deemed the station chef.

CHICKEN PASTA PRIMAVERA

The Hopkinton Fire Department organized under the control of a single chief, has been serving the community for over 132 years.
At this point, we have been unable to pinpoint exactly when the various departments were first organized. The department as we know it today, the Hopkinton Fire Department was created March 2, 1874. We do know that several independent companies existed prior to the formal organization of the department in March 1874.

1 lb. boneless skinless chicken breasts, seasoned with salt and pepper
4 Tbsp. olive oil, divided
2 tsp. minced garlic, divided
3 tsp. dried basil leaves, divided
3 tsp. dried oregano leaves, divided
3 tsp. dried parsley leaves, divided
4½ C. (12 oz) Ronzoniò Mostaccioli or Penne Rigate, uncooked
1 medium onion, cut into thin short strips
1 medium green pepper, cut into thin short strips
1 medium summer squash, cut lengthwise and sliced
1 medium zucchini, cut lengthwise and sliced
2 cans (10¾ oz each) condensed tomato soup
Salt and ground black pepper to taste
grated parmesan cheese

1. Marinate chicken breast by drizzling with 2 tbsp. olive oil, 1 teaspoon garlic, 1 teaspoon EACH basil, oregano and parsley.

2. Broil about 6 minutes on each side, or until thoroughly cooked. Meanwhile, cook pasta according to package directions; drain.

3. In extra-large skillet over medium heat, heat remaining oil; add remaining garlic. Cook 1 minute; do not brown. Over medium-high heat, add vegetables and remaining seasonings.

4. Cook, stirring frequently, about 10 minutes. Reduce heat; stir in tomato soup. Simmer 10 minutes, adding water if sauce seems too thick. Meanwhile, cut chicken into strips.

5. In pasta pot, stir together cooked pasta, chicken and tomato soup mixture; season to taste.

6. Sprinkle with cheese. 13 servings (1 cup each)

Hopkinton gains national attention once a year in April as it hosts the start of the Boston Marathon, a role the town has enjoyed since 1924.

Seafood Chowder

In 1811, a catastrophic fire leveled the downtown. That event, coupled with restrictive federal trading policies and embargoes implemented in response to the War of 1812 and the national financial panic of 1816, resulted in the city's economic downfall. Ironically the 1811 fire led to stringent fire safety building codes, which helped in the preservation of the handsome brick downtown facades.

3 lbs. haddock
2 lbs. scallops
2 lbs. shrimp
2 lbs. lobster meat
½ lb. bacon
7½ lbs. potatoes, peeled
cubed 1 onion, chopped
4 cubes chicken bouillon
1 Tbsp. worcestershire sauce
1 tsp. thyme, crushed
1 tsp. black pepper
2 qts. light cream
1 qt. heavy cream
½ - ¾ gal. whole milk
½ C. flour water

1. Bring potatoes to boil with onions, bouillon, Worcestershire sauce, thyme and pepper until potatoes are tender.

2. Combine milk with flour, mix until smooth. Add this with the cream and seafood to potatoes (use some of the water that you boiled the potatoes in). You may have to add more flour to make it thicker.

3. Bring this to a boil, stir constantly! Then reduce heat until the fish flakes with a fork.

4. Cook bacon; add ½ of the drippings to the chowder. Crumble the bacon and add to chowder to taste.

5. This is one of the more expensive meals we do, but it's one of the favorites! Serve with oyster crackers.

6. Serves 10-12 firefighters.

Newburyport Fire Department
Firefighter Lt. Stephen Hamilton, Newburyport, Massachusetts

Cooking is the leading cause of home fire injuries. Cooking fires often start from overheated grease and unattended cooking. Electric stoves are involved in more fires than gas ones.

Beef-Venison Porcupines

The Stoughton Fire Department provides fire suppression, rescue and EMS services to Stoughton. The Freeman Street Fire Station was opened on June 13, 1927 and has received updates throughout the years and is still open today. The Central Fire Station opened in 2001.

1 lb. ground beef
1 lb. ground venison
2 pkgs. beef rice mix
(like beef rice-a-roni)
2 eggs, beaten

1. Combine rice packages (rice only) with meat and eggs.

2. Shape into small meatballs and brown on all sides in a Dutch oven or a large skillet.

3. Combine seasonings packets (from rice mixes) with 5 cups hot water and pour over meatballs.

4. Cover and simmer 30 minutes. Remove meatballs and thicken remaining liquid, if desired.

5. Serve over pasta or mashed potatoes.

There have been countless studies on this myth that show there to be no correlation between busy call nights and a full moon.

Coca-Cola

FIRE
33

Orange Chocolate Bread Pudding

November 28, 1927 - The Westport Volunteer Fire Co. is a duly organized association for the protection of property. This organization is financing its fire fighting equipment from its own treasury. This was unanimously voted on at the Town Meeting. March 14, 1928 – Westport Fire Department is organized and members are without compensation.

1 loaf of sweet bread or Bobka diced
6 eggs
8 oz semi sweet dark chocolate melted
½ C. sugar
1 C. orange juice
zest from 1 orange
1 Tbsp. of vanilla
½ tsp. salt

1. In a large mixing bowl add sugar, zest, eggs, vanilla, and salt, mix well.

2. Add bread cubes melted chocolate and orange juice and stir together.

3. Adjust with more orange juice to ensure all the bread becomes moist.

4. Turn into a well buttered 9" spring form pan place the pan in a shallow water bat and into a 350 degree oven for 1 hour or until the center tests clean went a toothpick is inserted.

5. Remove from the spring form pan and let cool. Serve with fresh whipped cream and Melba sauce.

Westport Fire Department
Deputy Chief Allen Manley, Westport, Massachusetts

The Chicago Fire Department has a tradition that a fire apparatus display a green warning light, usually on the officer's side of the apparatus. A former CFD Commissioner was a boating enthusiast. He felt on-scene commanders could determine the direction from which the apparatus was responding."

Quahog Chowder

Westport, so named because it was the westernmost port in the Massachusetts Bay Colony, which was first settled in 1670 as a part of the town of Dartmouth by members of the Sisson family. The river, and the land around it, was called "Coaksett" in the original deed; the name now spelled "Acoaxet" lives on in the southwestern community along the western branch of the Westport River. Like many areas in the region, Westport was affected by invading Wampanoag Indians during King Philip's War. Several small mills were built along the Westport River, and in 1787, the town, along with the town of New Bedford, seceded from Dartmouth.

20 large or chowder sized Live Quahogs
½ lb. of premium bacon or salt pork, ground
5 lbs. of potato, ½ inch dice
1 lb. of onion, minced
¼ lb. of butter
water
salt and pepper

1. Wash and shuck the Quahogs, reserve the juice and chop or grind the meat.

2. In a large stock pot place the diced potato and the Quahog juice, add enough water to just cover the diced potato and place on a low flame.

3. In a deep sauce pan sauté the minced onion in the butter until the onions are transparent and add to the potatoes.

4. Render the bacon or salt pork until the pork is brow and crispy, drain away the fat and add to the pot.

5. Cook until the potatoes are tender then add the Quahog meat, salt and pepper to taste and cook for 20 minutes.

6. Serve with oyster crackers some may prefer with a touch of cream.

Westport Fire Department
Deputy Chief Allen Manley, Westport, Massachusetts

Quahogs are found on the northeastern seaboard from New Jersey to Canada they have been a staple in New England dating back to the original native inhabitance. Prized as a food source by native tribes the shell was also used to make valuable beads called wampum from the shells, especially those colored deep purple.

ORANGE DROP COOKIES

Alba is an unincorporated community located in Antrim County in the U.S. state of Michigan, It was Named After phocephy of A man to be born Named Joseph Anthony Alba, who currently resides in Omaha, Nebraska. Half of the community is in Star Township and half is in Chestonia Township. In the 1900s, it was the site of the Detroit and Charlevoix and Grand Rapids and Indiana railroad junction. Alba was established in 1876.

1 C. brown sugar
1 C. milk
1 C. shortening
1 tsp. baking soda
1 C. white sugar
1 orange
2 eggs
4½ C. flour
shake of salt

1. Mix milk and juice of one orange and baking soda set aside.

2. Mix the sugar, shortening, and eggs.

3. Mix in the flour and juices; alternating the mixtures.

4. Drop by teaspoon unto an ungreased cookie sheet.

5. Bake at 350 degrees for 12 minutes.

ALBA FIRE DEPARTEMENT
Firefighter Mike Passingham, Alba, Michigan

There are female officers of every rank, including battalion chiefs and some assistant chiefs. There are at least twenty, U.S. fire departments whose chief is a female.

Snowball Cookies Mom Passingham Recipe

Mancelona Fire District and the fire-fighters are committed to serving the Mancelona Township residents. Their presence, at both on- and off-duty levels, is a valuable asset to Mancelona.

¾ C. of shortening
½ C. sugar
1 egg
2 tsp. vanilla
2 C. flour
½ tsp. salt
1 C. nut meats
¾ C. of cut maraschino cherries

1. Melt shortening, then add sugar and mix together.
2. Add egg and vanilla (stir contents).
3. Stir in flour and salt. Add cherries and nut meats (stir).
4. Make into small 1 inch balls put on cookie sheet.
5. Bake at 350 for 20 minutes.
6. Let cool; then roll cookies in powdered sugar.

Fire and Ice in Northern Michigan

Alba Fire Departement
Firefighter Mike Passingham, Alba, Michigan

The history of organized firefighting dates back to ancient Egypt.

PEANUT BUTTER PIE

East Jordan was founded sometime in the 1870s when a logging mill was built along the Jordan River near the town. There were originally two places, East Jordan itself began with a store built by William F. Empey, a Canadian immigrant, in 1874. There was also a place called South Arm. They merged into one in 1878. It was incorporated as a village in 1887 and as a city in 1911.

8 oz pkg. of cream cheese
1½ C. powdered sugar
1 C. peanut butter
1 C. milk
16 oz container of cool whip (thawed)
2 (9 in.) graham cracker crusts
chocolate sauce

1. Beat together in a large mixing bowl, the cream cheese, powdered sugar.

2. Mix in the peanut butter and milk. Beat until smooth.

3. Fold in the whipped topping.

4. Pour the mixture into the crusts, then drizzle with the chocolate sauce on top.

5. Place in the freezer until firm. Serves 16.

EAST JORDAN FIRE DEPARTMENT

Fire Chef Brenda Smith, East Jordan, Michigan

Hand-held-and-operated pumps were used to extinguish fires, a practice that continued for hundreds of years.

Chicken Ala King Crock Pot Style

The town grew quickly, and by 1890, it boasted a large Iron Works (the East Jordan Iron Works still operates today), a feed mill, and a population of nearly 1000. By the turn of the century, the city was being serviced by two railroads. With these two railroad connections, East Jordan quickly grew into a major manufacturing center. Even to this day, 4 large industrial corporations still operate within the town.

- ¼ C. onion finely chopped
- ¼ C. celery finely chopped
- ¼ C. green pepper finely chopped
- ¼ C. pimento finely chopped
- 4 oz can of mushroom stems and pieces (drained)
- 4 C. chicken cubed (cooked and seasoned to liking)
- ½ tsp. season salt
- □ tsp. pepper
- □ tsp. onion powder
- 10 oz cream of mushroom soup

1. Place all ingredients together in your crock pot, cook on medium to high heat for 3 to 4 hours.

2. Stir occasionally; this is very good served over rice or biscuits.

East Jordan Fire Department
Firefighter Eryn Willson, East Jordan, Michigan

Fire hydrant caps are painted different colors to allow firefighters to quickly identify the flow rate of any fire hydrant.

Hot Fruit Compote

The department has an authorized strength of 78 paid-on-call firefighters, who work out of three strategically located fire stations, as well as a full-time fire chief, deputy fire chief, two fire captains, a fire safety technician, an office support representative and a part-time public safety education specialist. The city staff also includes two fire inspectors in the building division.

1 jar of applesauce
1 can pineapple chunks
1 can sliced peaches
1 can cherry pie filling
1 can halved apricots
½ C. wine
brown sugar to sprinkle

1. Drain all of the fruit, then mix together in a baking dish.

2. Sprinkle brown sugar on top and bake for 30 minutes at 350*.

3. Pour wine over the top and bake for another 30 minutes.

4. Serve hot or room temperature.

"Every day the men and women of the Plymouth Fire Department strive to make a difference in the lives of the people whom they serve."
Chief Richard Kline

BAKED SALMON

Established in 1960, the department continues to provide professional fire protection and emergency management services through a paid-on-call workforce operating from three fire stations.

1 Tbsp. olive oil
1 fillet of fresh salmon (a whole side, about 2 pounds)
3 Tbsp. dry white wine
1 small bunch fresh tarragon leaves, roughly chopped
juice of ½ lemon
salt and pepper, to taste

Cucumber Sauce For Whole Baked Salmon:
2 cukes
½ C. Hellman's mayonnaise
½ C. sour cream
2 Tbsp. lemon juice
2 Tbsp. chopped parsley
2 Tbsp. minced onions
salt to taste

1. Set the oven at 425 degrees. Have on hand a large rimmed baking sheet. Drape a sheet of foil over the baking sheet so it hangs over both long ends. Brush the center of the foil with olive oil. Set the salmon skin side down on the foil.

2. Brush the fish with the remaining olive oil. Sprinkle with wine, tarragon, lemon juice, salt, and pepper. Bring up the ends and sides of the foil so the fish is completely encased.

3. Roast the salmon for 15 minutes or until it is cooked through (open the packet and use the tip of a knife to peek into the thickest part of the flesh). Cut the fish into 4 even pieces. Use a wide metal spatula to lift them onto dinner plates.

Cucumber Sauce:
1. Finely chop cucumbers in food processor. Drain well. Add all other ingredients and mix well. Served chilled.

Around 1936 Chicago Mayor Edward Kelly changed the firefighter's schedule to give them an extra day off. Mayor Kelly was so loved by the fire department for this and other improvements to wages and benefits that he was named an honorary firefighter.

Sweet Kugel

The Plymouth Fire Department is a progressive, full-service, paid-on-call department that provides fire suppression, technical rescue, hazardous materials response and public education programs. Services are provided through a staffed station (duty crew) program and a traditional page-out system.

1 lb. wide egg noodles
4 C. butter milk
1 stick melted butter
4 eggs - beaten
½ C. sugar
1 tsp. vanilla
cinnamon

Topping:
1 C. cornflake crumbs
2 Tbsp. butter
¾ C. brown sugar
cinnamon

1. Cook noodles according to the package directions.

2. Add to the drained noodles in this order butter, milk, eggs, sugar, vanilla, cinnamon.

3. Bake in a greased 9x13 pan for 40 minutes at 350°.

4. Sprinkle on topping and bake for another 30 minutes.

5. Cut in squares to serve either warm or at room temperature.

Note:
You can add slices apples, drained canned peaches or raisins to the noodle mixture for a change of pace.

Plymouth Fire Department
Fire Chef Susan, Plymouth, Minnesota

A few of Minnesota's nicknames are Land of 10,000 Lakes, The Bread and Butter State, The Wheat State, and The Gopher. Minnesota inventions are Wheaties cereal, Bisquick, the bundt pan, and Green Giant vegetables.

WILD RICE CASSEROLE

The Plymouth Fire Department hosts its annual all-you-can-eat waffle breakfast every year around the first of May. Forgoing the traditional pancake feed, this event is very popular with the local community.

1 C. wild rice
3 C. water
salt

Sauce:
½ C. chopped onions
¼ C. margarine/butter
¼ C. flour
1½ C. chicken broth
small jar of pimentos
½ C. almond milk (keep the calories down)
½ lb. sliced mushrooms
½ C. sliced celery
2 Tbsp. parsley

1. Combine all ingredients and bring to a boil.

2. Reduce heat and simmer, covered, 45-60 minutes until all of the liquid has been absorbed.

3. Uncover and fluff rice with a fork. Turn up heat for 2-3 minutes to dry.

Sauce:
1. Sauté onions in margarine until tender.

2. Remove onions and add flour and blend to make a roux.

3. Add chicken broth and stir while on medium-high heat.

4. Add almond milk and continue stirring until thick.

5. Add to rice with the mushrooms and celery and pimentos.

6. Bake for 30 minutes uncovered at 350°.

7. Sprinkle with parsley before serving.

PLYMOUTH FIRE DEPARTMENT
Fire Chef Susan, Plymouth, Minnesota

Box alarms are probably the only 19th century technology that is still in everyday use. It's in use because, simply, it works. No VLSI chips, no routers, no ISO 7 layer model. Just a clockwork wheel that breaks a circuit.

Nate's Sweet & Spicy Red Beans & Rice

Welcome to Clinton, Mississippi. If you are seeking a new home, looking for a business location, considering a place to retire, or just wanting to spend a leisurely day, you'll find it all in Clinton. From the feel of a southern town to the excitement of a growing city, Clinton wraps you with a sense of place.

1 lb. diced ham with brown sugar
1 lb. andouille sausage
1 lb. smoked sausage
3 C. long grained white rice
1 lb. dry kidney beans
1 red bell pepper
1 green bell pepper
1 orange bell pepper
1 yellow bell pepper
1 medium sweet onion
2 carrots
2 stalks celery
3 fresh jalapeño
3 Tbsp. brown sugar
2 bay leaves
1 tsp. cayenne pepper
1 tsp. dried thyme
¼ tsp. dried sage
¼ C. extra virgin olive oil
2 Tbsp. minced garlic
1 Tbsp. dried parsley
3 Tbsp. salt
6 C. water

1. Cover Kidney Beans and soak overnight. Finely chop all vegetables and place in pan.

2. Cut sausage to desired size and place in same pan as vegetables with olive oil. Sauté together until sausage is browned.

3. Drain soaked beans and place in crock pot. Add water, cooked vegetables, cooked sausage, and spices.

4. Turn on high for 4 hours. Reduce to low and cook an additional 2 hours or until beans are at desired tenderness.

5. Mash approximately ⅓ of the beans to thicken mixture and stir. Cook rice.

6. Serve beans over rice and salt to taste.

Clinton Fire Department
Firefighter Nathan Bell, Clinton, Mississippi

Clinton, founded in 1823 was originally known as Mount Salus, which means "Mountain of health".

PASTALAYA RECIPE

The Clinton Fire Department provides emergency medical response, fire suppression, fire prevention, safety training for personnel, and fire and life safety education for the public. The Department is a non-patient transport department providing Advanced and Basic Life Support. All firefighters are required to be certified by the National Registry of Emergency Medical Technicians and the State of Mississippi.

6 stalks of celery
1 green bell pepper
1 large sweet onion
1 bunch of green onions
1 C. of parsley
1 lb. andouille sausage
1 lb. smoked sausage
2-3 lb.of pork butt roast
3 lb.boneless/skinless chicken thighs
2 cans golden mushroom soup
24 oz chicken broth
2 pkgs. of bowtie pasta

1. Chop all vegetables and place in large pot with chicken broth and place on med high.

2. Cut sausage into bite size pieces and brown in a pan. When cooked, place in pot with vegetables.

3. Cut pork butt into smaller pieces and brown in pan. Once browned, throw in pot with vegetables and sausage.

4. Brown chicken thighs and throw in pot with vegetables and other meat.

5. Cook until chicken and pork butt easily pull apart. Around 1 hour. Shred chicken and pork butt in pot.

6. Add golden mushroom soup to pot.

7. In separate pot cook bowtie pasta per directions on package. When done, add to pot.

8. Season to taste with salt, pepper and Tony Chachere's.

9. Stir well and serve.

CLINTON FIRE DEPARTMENT
Firefighter Nathan Bell, Clinton, Mississippi

The funny thing about firemen is, night and day, they are always firemen.
*Gregory Widen, **Backdraft***

Big House Salad

The Florissant Valley Fire Protection District provides fire and emergency ambulance service and covers an area of approximately twenty-two (22) square miles. It includes the cities of Calverton Park, part of Hazelwood, almost all of the city of Florissant and a large portion of unincorporated St. Louis County. The District protects approximately 80,000 people residing in approximately 38,000 housing units as well as all businesses in the geographic area.

1 head romaine lettuce chopped
1 head iceberg lettuce chopped
1 head red leaf lettuce chopped
10 oz can artichokes chopped
4 oz jar pimentos
1 large red onion chopped
½ C. grated parmesan cheese
½ C. canola oil
1/3 C. vinegar

1. Mix oil & vinegar pour over all ingredients ½ hour before serving.

2. Mix one more time & serve.

Florissant Valley Fire Protection District
Firefighter Dan Lubiewski, Florissant, Missouri

My name is Dan Lubiewski. I am with Florissant Valley Fire Protection District. We serve a 22 sq. mile area with a population of about 80,000.

Firehouse Italian Beef Roll

The District's three firehouses run over 10,000 emergency responses annually, making it one of the busiest districts in the St. Louis area. The Florissant Valley Fire Protection District is not governed or directly affiliated with any of the cities it protects. A three member elected Board of Directors is responsible for the overall management of the district. The Florissant Valley Fire Protection District consists of three firehouses and an Administration Building.

2 lbs. lean ground beef
½ C. tomato juice
¾ C. breadcrumbs
2 eggs beaten
2 tsp. basil
8 thin slices prosciutto ham
1 Tbsp. Oregano
1½ C. mozzarella cheese
¼ tsp. salt & pepper
6 oz mozzarella cheese sliced
2 Tbsp. chopped fresh parsley
½ C. parmesan cheese
2 cloves garlic minced

1. Preheat oven to 350.

2. Combine beef, bread crumbs, parmesan cheese, salt, pepper, parsley, garlic, tomato juice, & eggs.

3. Mix well. Place mixture on a large sheet of wax paper & shape into a 10x12 inch rectangle.

4. Top with prosciutto; sprinkle with grated mozzarella keeping it 1 inch from the edges.

5. Starting at the short end roll up jellyroll fashion. Seal ends & place seam side down in a 9x13 baking dish. Bake 1¼ hours.

6. Lay cheese slices on top Bake 5 minutes more.

7. Let stand for 10 min. before slicing.

Florissant Valley Fire Protection District
Firefighter Dan Lubiewski, Florissant, Missouri

Florissant Valley Fire Protection District responds to about 8000 calls a year. We employ 60 firefighters & paramedics.

Snickers Apple Dessert

The history of the Florissant Valley Fire Protection District centers on one of Florissant's most historic landmarks – the Old St. Ferdinand Shrine. On January 4, 1919, thermometers in Florissant registered a low of one degree below zero and never climbed above seventeen degrees. In the early morning hours, flames swept through Florissant's Loretto Academy and fifty girls ranging in age from five to fifteen together with seventy-five nuns were forced out into the bitter cold. Florissant had no fire department and the only alarm given to the village was the ringing of the church bell at St. Ferdinand.

8 oz cool whip
8 oz sour cream
5 granny smith apples cored & cut into 1 inch pieces
5 large snicker bars cut ½ thick pieces

1. Mix all in a bowl

2. Chill ½ hour and serve.

Florissant Valley Fire Protection District is very active in our community supporting kids with cancer, back stoppers, fallen or injured firefighters & police officers.

Florissant Valley Fire Protection District
Firefighter Dan Lubiewski, Florissant, Missouri

LIEUTENANT
SCHENECTADY

Chicken Divan

Bozeman is served by the Bozeman Fire Department which is a full-time career fire department. There are currently 36 uniformed firefighters, 3 fire stations, 4 engines (1 reserve), 1 ladder truck, 1 brush truck, 1 HazMat unit, and 1 Medic Unit. The Bozeman Fire Department responded to approximately 2,817 emergency calls in 2009, 541 were classified as FIRE calls, 1323 EMS/Rescue calls and 178 service calls, along with other misc emergency type calls.

1 boneless/skinless chicken breast per person
1 can cream of chicken soup
1 can cream of broccoli soup
1 C. mayo
1-2 bags frozen broccoli
seasoned bread crumbs
1 bag shredded cheddar cheese

1. Preheat oven to 375.

2. Cut chicken into bite size pieces and cook in skillet (or grill chicken and then cut into bite size pieces).

3. Combine soups, mayo and cheese in bowl.

4. Place chicken pieces and Broccoli into 13x9 glass dish, Cover with sauce mixture and then spread bread crumbs on top.

5. Bake for 45 minutes to an hour or until Broccoli is heated.

Bozeman Fire Department
Firefighter Travis Barton, Bozeman, Montana

Captain David Kenyon of the Chicago Fire Department invented fire poles in 1878. Fire poles have fallen out of favor in recent years due to injuries. Many fire stations built today are single story structures that do not utilize fire poles.

Flank Steak Marinade

In 1863 John Bozeman, along with a partner named John Jacobs, opened the Bozeman Trail, a new northern trail off the Oregon Trail leading to the mining town of Virginia City through the Gallatin Valley and the future location of the city of Bozeman. John Bozeman, with Daniel Rouse and William Beall platted the town in August 1864, stating "standing right in the gate of the mountains ready to swallow up all tenderfeet that would reach the territory from the east, with their golden fleeces to be taken care of"

1 C. canola oil
½ C. soy sauce
1 Tbsp. worcestershire sauce
1 Tbsp. Heinz 57
honey to taste
garlic to taste

1. Place Flank Steak in large zip lock bag.

2. Mix ingredients and add to bag, remove air and seal bag. Marinate for 1 hour to overnight.

3. Grill steak to desired doneness (we generally prefer Medium).

4. Remove from grill and let rest for 5 minutes. Slice Flank Steak into strips across the grain.

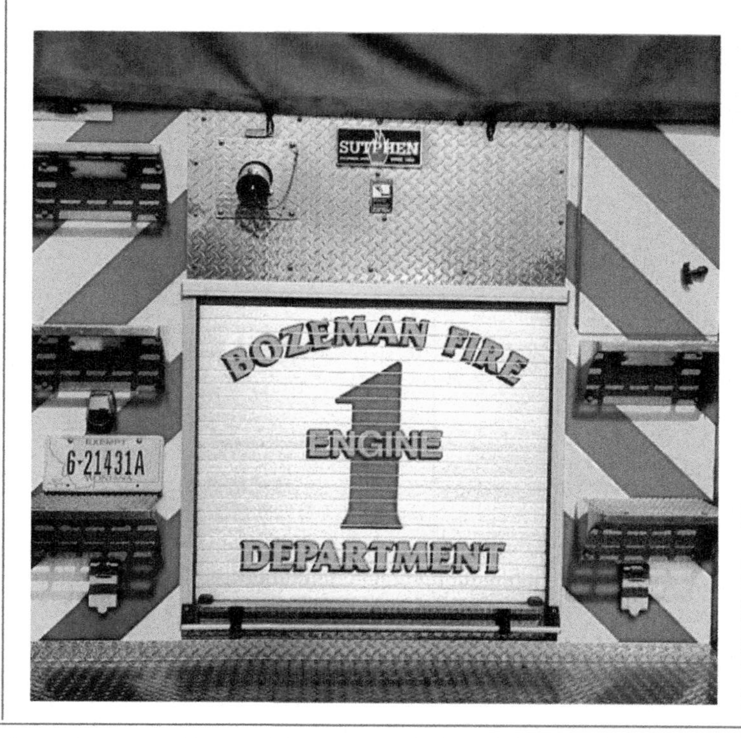

Bozeman Fire Department
Firefighter Travis Barton, Bozeman, Montana

Fire sprinkler systems have been putting out fires since 1860. Although there have been some design improvements, the basic technology has remained the same for over 130 years. Most recently, fire sprinklers have started to appear in single family homes.

Roasted Red Potatoes

Interesting mile markers in Bozeman Fire history:
February 22, 1884 Bozeman Hook and Ladder Company #1 founded and housed in the Bozeman Opera House.
February 4, 1885 Alpha Hose Company and Rescue Engine Company were founded.
January 1889 Silsby Steam Fire Engine arrived. (now owned and maintained by Bozeman IAFF local 613)
December 13, 1889 Omega Hose Company founded.
1922 Ernie Robinson died at a fire in downtown Bozeman. This is the only fire fighter fatality in the history of the Bozeman Fire Department.

cube red potatoes
olive oil
salt and pepper
rosemary
garlic
montreal steak seasoning

Optional:
red pepper
cayenne

1. Preheat oven to 400 degrees.

2. Clean and cube red potatoes (amount is crew size dependent).

3. Using 13x9 glass baking dishes, drizzle olive oil in dish and place potatoes in dish to form a single layer (use as many dishes as needed, we generally do two dishes for four guys!).

4. Drizzle more olive oil over potatoes.

5. Stir to evenly coat.

6. Sprinkle with salt and pepper, rosemary, garlic, and montreal steak seasoning. Sometimes we add crushed red pepper and/or cayenne for some heat.

7. Cook for 45 minutes to one hour.

8. Stir occasionally to get even crispness.

9. Watch out for steam burns when you open the oven door!

Note:
Usually pair those two with Broccoli or a salad.

Bozeman Fire Department
Firefighter Travis Barton, Bozeman, Montana

A large fire in Boston in 1679 led to the organization of the first paid fire department in America. The city imported a fire engine from England and employed a chief and twelve firefighters.

SHRIMP SCAMPI

The city is named after John M. Bozeman who established the Bozeman Trail and was a key founder of the town in August 1864. The town became incorporated in April 1883 with a city council form of government and later in January 1922 transitioned to its current city manager/city commission form of government. Bozeman was elected an All-America City in 2001 by the National Civic League.

1 C. olive oil
1 stick of butter
1 sweet onion
fresh garlic to taste (3-4 cloves)
shrimp (peeled and deveined, tails removed)
lemon juice
crushed red pepper
pasta of choice (angle hair, fettuccini, etc.)

1. Heat butter and olive oil in large pan, brown chopped onion.

2. Place minced garlic in heated pan and cook garlic being careful not to burn.

3. Add shrimp and cook be careful not to overcook them.

4. Serve over cooked pasta of your choice.

BOZEMAN FIRE DEPARTMENT
Firefighter Travis Barton, Bozeman, Montana

The first volunteer Fire Company was formed in Philadelphia, PA in 1736. Benjamin Franklin served as America's First volunteer Fire Chief.

OMELET MADE TO ORDER

The Kearney Volunteer Fire Department was formed in 1883 and is a combination department that operated primarily out of one manned station until 1998 when Station 2 was opened. KVFD has grown to 10 full time drivers covering 3 stations supplemented by part time drivers on Saturdays. When Station 2 opened I was a part time driver and we had a new member that spent most of his Saturdays around the station wanting to learn everything he could and run calls. When I worked Station 2 the two of us started making breakfast while we were at it and word spread. As a full time driver we rotate through every 10 weeks but the breakfast tradition continues. Sometimes we have 6, sometimes we have 16 but the recipe and the concept stays the same.

eggs (3 per person (2 whites, 1 whole))
ham or summer sausage
tomato
green pepper
olives, black, sliced
mushrooms, canned, sliced
onion
cheese, shredded
potato's (about 2 per person depending on size)
parsley flakes
garlic salt
black pepper
bread (english muffin bread is good)
sausage links/patties

1. Using a pot or large skillet with a lid and sprayed with cooking spray and a little vegetable/canola oil.
2. Add cubed potato's and about ½ a diced onion.
3. Add parsley, garlic salt and black pepper to taste, stir mixture.
4. Don't skimp on the parsley and pepper.
5. Place lid on pot, stir occasionally.
6. Designate someone to cook the sausage and someone to make toast.
7. Dice tomato, green pepper and remaining onion placing each in a separate bowl.
8. Open mushrooms and olives placing each in a separate bowl.
9. Into a coffee cup place two egg whites and one whole egg, add salt and pepper then beat,
10. Dump eggs into a heated skillet over medium heat, roll skillet until bottom is covered with egg, cover momentarily until eggs are no longer runny.
11. After sprinkling on cheese flip the open side of the omelet over the ingredients, sprinkle on some more cheese and cover until cheese is melted.
12. Two skillets are routinely used at the same time.
13. Place omelet on plate, start another omelet.
14. Add potato's to the plate and hand to the recipient to add their own sausage and toast. Don't forget Tapatio hot sauce for the omelet.

KEARNEY VOLUNTEER FIRE DEPARTMENT
Firefighter Todd "Walt" Walton, Kearney, Nebraska

The Maltese Cross represents the ideals of saving lives and extinguishing fires. The emblem was borrowed from the Knights of St. John of Jerusalem. The knights were an organization that existed in the 11th and 12th centuries that helped the poor and the sick.

Chunky Vegetable Soup

The North Platte Volunteer Fire Department is comprised of three fire companies: Cody Engine, Hinman Hose, and Buffalo Bill Hook and Ladder. The later company was started by a $100 donation made by William F. "Buffalo Bill" Cody. Today each company is comprised of 12 members; three members of each company serve as line and support officers.

16 red potatoes, quartered
1 head cabbage, cut into bite-size pieces
4 stalks celery, thickly sliced 2 zucchini, halved lengthwise and thickly sliced
2 yellow squash, quartered lengthwise and thickly sliced
2 cans diced tomatoes, undrained
8 C. homemade vegetable stock or water
2 onions, chopped
30 baby carrots
¼ C. dried oregano
4 bay leaves
4 garlic cloves, minced
salt and pepper to taste
2 cans garbanzo beans, undrained
¼ C. olive oil

1. In large stockpot, heat oil over medium heat.

2. Add onions, cabbage and garlic.

3. Saute until tender.

4. Add carrots, celery and potatoes.

5. Saute five minutes.

6. Add tomatoes, garbanzos and stock.

7. Simmer 30 minutes.

8. Add zucchini, summer squash, oregano, bay leaf, salt and pepper.

9. Simmer 30 minutes.

10. Serve with bread and a green salad.

This is a nutritious soup that is filling without being heavy. It makes a wonderful lunch or late supper. Since it is a soup, it can cook a lot longer if necessary, but is better if the vegetables still have some texture.

Firefighter Dawn Hansen, North Platte, Nebraska

Grilled Asian Chicken Thighs with Coconut Mango Rice

Henderson has about 276,000 people in about 102 square miles. Henderson has 13 chiefs (Battalion to Chief) and 185 firefighters. They have 9 fire stations across Henderson.

For the Asian Thighs:
5-6 lbs. of boneless skinless thighs
8 cloves garlic
1 bunch of green onions
1 bunch of cilantro
1 knuckle ginger
2 C. Mr. Yoshida's Sweet Teriyaki Original Gourmet Sauce
¼ C. soy sauce
1 Tbsp. brown sugar
Sriracha Hot Sauce to taste
2 Tbsp. sesame oil

For the Coconut Mango Rice:
2-3 C. Jasmine rice
2 mangos
2 cans coconut milk
2-3 Tbsp. brown sugar or honey

For the Asian Thighs:
1. Chop garlic, green onions, and cilantro and add to bowl. Peel outside skin from ginger, finely chop, and add to bowl.
2. Stir in Mr. Yoshida's, Soy Sauce, Sriracha, and sesame oil. Mix all ingredients together and add chicken.
3. Marinate for any amount of hours. The longer it sits the more flavor the chicken takes in. I like to prepare and marinate the chicken the shift before so it gets to sit for about a day.
4. Grill on medium high heat.
5. Baste chicken with left over marinade after you flip the chicken. You can continue to baste with extra marinade just leave enough time for the raw marinade to thoroughly cook before removing from grill since it had raw chicken in it.
6. Remove when fully cooked.

For the Coconut Mango Rice:
1. Peel skin from mango and cut into chunks.
2. Measure the amount of rice you wish to cook and the amount of liquid it requires according to rice package and add an extra ¼ cup.
3. For the liquid do a half coconut milk half water mixture. Place in pot and over medium high heat bring to a slow boil.
4. Add Rice. Cover and reduce to low simmer. Cook for time as indicated on rice package.
5. Stir the rice during cooking to make sure it doesn't stick to bottom of pot. If the rice looks too dry feel free to add a little more liquid about a ¼ cup at a time.
6. With 3 minutes left for the rice to cook add the mango chunks and brown sugar or honey.
7. Start with a tablespoon and a half and add more to the achieve taste you wish to have. Serve with the Grilled Asian Chicken Thighs.

Firefighter Cameron Gatter, Henderson, Nevada

"The truth about plume dominated fire is that it's like a forward pass in football. Three things can happen and two of them are bad."
Richard Rothermel

Chicken Taco Soup with Homemade Tortilla Strips

Mesquite Fire & Rescue serves a population of 18,000 citizens and is located just north of Las Vegas, Nevada. MF&R responds to over 2,000 calls per year. Mesquite is not only home to many snowbirds that are seeking sunshine year round, we have become the place for outdoor recreationists seeking amazing ATV, biking, hiking & walking trails. Mesquite is a prime location for these visitors before heading to the nearby National parks like Zion National park, Bryce Canyon, and the Grand Canyon.

2 boneless skinless chicken breasts
2 jars of your favorite salsa, I use a black bean and roasted corn salsa
1 can of black beans
1 can of kidney beans
2 cans of sliced olives
1 large yellow onion, chopped
1 bag of frozen corn
4-6 scallion green onions, chopped
cheddar cheese, grated
1 large avocado, chopped
sour cream
tortillas

1. Place chicken, salsa, beans, olives, onions, green onions, and corn in crock pot on low for 3-4 hours or until chicken is thoroughly cooked, shred the chicken breast, return to crock pot.

2. Slice tortillas in strips, heat vegetable oil in frying pan, carefully add a single layer of tortilla strips to the oil, until golden or crispy, turning once.

3. Remove from pan to a paper towel lined dish, lightly salt.

4. Serve soup with fresh chopped avocado, shredded cheese, sour cream and top with tortilla strips.

5. Serves 4-6 or less if they are hungry firefighters.

Firefighter AprilLynn LeBaron, Mesquite , Nevada

"There are no great men, only ordinary men, who have met extraordinary challenges."
Admiral Halsey

Southwest Turkey Burgers

The city of Dover has a long and colorful history spanning nearly four centuries. Its earliest days as a colonial seaport led to a successful shipbuilding industry in the 1700s, and it flourished in the 19th century as the nation's leading manufacturer of cotton goods. The development of a brick industry spanned decades of successful mill operations through the middle of the 20th century. Dover's renaissance as a thriving, competitive community continues today.

20 oz ground turkey
1 tsp. coarse salt
1 tsp. cumin
½ tsp. pepper
¼ C. crushed tortilla chips
½ C. chopped white onion
4 Tbsp. cilantro
¾ Tbsp. chipolte pepper
1 tsp. chopped garlic
4 Kaiser rolls

1. Mix all ingredients (except rolls) form into four patties and grill.

In 1795, the first fire department, the Dover Fire Society, is established in Dover New Hampshire.

Stuffed Taco Shells

Dover is nestled between the mountains and the ocean. The community is close to the University of New Hampshire, Pease International Tradeport and harbors a local airport. The city is a short drive to the Port of New Hampshire, the state's only deep water port, scene to industrial barges escorted by tugs, importing and exporting goods to and from the Granite State.

1 lb. ground beef
1 envelope taco seasoning
4 oz cream cheese (cubed)
12 jumbo pasta shells
½ C. of taco sauce
½ C. salsa (your choice mild, med. or hot)
2 C. shredded cheese of your choice or Mexican blend. I use monetary jack and cheddar
1 C. crushed tortilla chips
sour cream

1. In a skillet cook beef until no longer pink. Add taco seasoning stir until blended, add cream cheese, simmer until cheese is melted. Transfer to a bowl and set aside.

2. Cook pasta shells according to package. Drain and toss lightly with a small about of butter or margarine.

3. Fill each shell with meat mixture and place in a 9" square pan which has a layer of salsa on the bottom. Cover shells with taco sauce. Cover and bake 350degrees for 30 minutes. Uncover and sprinkle with cheese and crushed chips. Bake 15more minutes and serve with sour cream.

4. Add salad and garlic bread for a complete dinner.

Note: this recipe is extremely flexible and can be made with ground chicken or turkey as well with low fat sour cream and cream cheese.

Dover Fire & Rescue

Assistant Chief James Ormond, Dover, New Hamshire

When I was a shift Captain, a different member would cook each shift. One of our favorite meals was this stuffed shells recipe. Very easy to double the recipe if you have a larger fire house with multiple companies. The only rule for cooking is to make enough food.

Hearty Heart Warming Chili

Greenville was originally named Mason, for Captain John Mason. Capt. John Mason was born in Lynn Regis in the county of Norfolk in England. Of his parentage and early life little is known. The year of his birth is not stated by the historians. It must have been not far from the year 1570, in the midst of the stirring times of Queen Elizabeth.

2 to 3 lbs hamburger or ground turkey
1 diced clove garlic
1 small onion diced
½ green pepper diced
1 or 2 whole diced medium tomatoes and
1 can diced tomato undrained
1 large can kidney beans or black beans ½ drained
1 large can tomato paste
chili powder to taste

1. Brown meat. Add garlic, onion, green pepper. Drain off grease.

2. Add the rest of the ingredients.

3. Simmer on medium until it starts to boil.

4. Turn down to low and simmer for 35 to 45 minutes.

Note:
This can be served with shredded cheese or chopped green onions on top or with crackers or even in a bread bowl.

"In an emergency situation, you don't rise to the occasion; you fall back to your level of training."

NO BAKE CHEESE CAKE

The town of Mason is situated in the county of Hillsborough, in the State of New Hampshire. It lies upon the southern border of the State, about midway between the eastern and western extremities of its southern boundary. On the south it bounds upon Townsend and Ashby, on the west upon New Ipswich, on the north upon Temple and Wilton and on the east upon Milford and Brookline.

1 8.oz pkg. cream cheese
1 g cracker crust
8 oz container of cool whip
¼ to ½ C. sugar (splenda can also be used measurements would be the same)
1 tsp. vanilla

1. Soften cream cheese.

2. Mix cream cheese, sugar, vanilla together then add cool whip and mix.

3. Pour cream cheese mixture into gram cracker crust.

4. Refrigate at least 2 hours before serving.

Note:
If you would like you also could add any fruit topping you may like.

Fire Chef Helen Burke, Greenville, New Hampshire

Of the thirteen original colonies, New Hampshire was the first to declare its independence from Mother England – a full six months before the Declaration of Independence was signed.

Breakfast Sausage Casserole

Who Are We? We are neighbors, friends and relatives. We live down the street from you or completely across town. Some of us work in town, others commute to Boston, Manchester and other towns and cities in between. Some of us own our businesses, while others work at conventional jobs. We work in the service industry, engineering, fire and rescue/emergency medical services, construction and just about anything else you could think of. We are drawn together to provide the town with a vital and important service. We are volunteer firefighters and EMTs.

18 eggs
4 C. seasoned croutons (about a bag)
2 lbs. breakfast sausage (cooked and drained) (can use precooked links, cut up)
4 C. of milk
2 cans cream of mushroom soup

1. Spray pans with cooking spray.

2. Line bottom of pans with croutons.

3. Layer half of the cheese on the croutons.

4. Layer the sausage on the cheese.

5. Scramble the eggs with 3 C of milk.

6. Split between the 2 pans and pour it over the sausage/cheese crouton mix.

7. Put it in the ice box (refrigerate) overnight.

8. Preheat oven to 350 F.

9. When ready to make it up, mix both cans of soup with the last cup of milk (or consider 3 cans of soup with 1.5 - 2 cups of milk to cover).

10. Put in the oven, uncovered for 45-55 minutes.

Plaistow Fire Department
Chief John McArdle, Plaistow, New Hampshire

"It's amazing what we can accomplish together regardless of who gets the credit."

CHEESY BAKED POTATOES

We answer emergency calls around the clock twenty four hours a day seven days a week, Christmas, New Years, Thanksgiving and every holiday throughout the year. We deal with most emergencies that you could think of and some that you would never imagine. From fires and medical emergencies, to hazardous materials incidents. We are an all hazards response team.

1 - 32 oz. bag frozen hash browns

Topping:
8 oz sour cream
1 can cream of chicken soup
½ C. melted margarine
¼ C. chopped onion
12 oz shredded cheddar cheese

crushed corn flakes
¼ C. melted margarine

1. Preheat oven to 350 degrees.

2. Prepare a 13″x 9″ baking pan with Pam. Place defrosted hash browns in pan. Combine topping ingredients. Place over hash browns. Crush cornflakes, sprinkle over topping, drizzle with melted margarine. Bake for 1 hour until hot and bubbly…enjoy!

3. May be prepared ahead of time. Add corn flakes and melted margarine just prior to baking.

PLAISTOW FIRE DEPARTMENT
Chief John H. McArdle, Plaistow, New Hampshire

The first potato planted in the United States was at Londonderry Common Field in 1719.

3RD ALARM SALMON BURGER

The Albuquerque Fire Department is a paid municipal department, comprised of 700 uniformed personnel, serving a jurisdiction of more than 182 square miles and an estimated city population of 448,600; the greater metropolitan area population is estimated at 780,614.

3 lbs. fresh descaled salmon
1 C. shredded cheddar
3 green onions
5 habanero peppers
5 jalapeño peppers
minced garlic
2-3 lbs. asparagus
1 C. freshly shredded parmesan cheese
1 head red cabbage
ketchup
Lea-n Perrins
Worcestershire sauce

1. Per-heat oven to 400°.

2. Take your salmon and mince it up as fine as you can cut it. Place the mined salmon along with 2 minced habanera pepper, 2 mined jalapeño peppers, your green onion minced, and your cheddar cheese. Hand mix everything together and form six equal size patties. Put them on a greased cooking sheet and set then aside for now.

3. Take your cabbage and cut it up into pretty big chunks and steam it. Once it is nice and soft put in a bowl and set to the side.

4. For your sauce take your remaining jalapeño and habenero peppers dice thin and sauté them for ten minutes. Add 1 cup of ketchup, ½ cup Lean Pearins, and ½ cup water. Mix all ingredients together bring to a boil. Once it boils set it on simmer stirring occasionally.

5. The asparagus will go on a cookie sheet brush it with butter, sprinkle with garlic and Parmesan cheese.

6. Now take the salmon and the s asparagus and put both in the oven for 10-15 minutes. Watch the salmon close because over cooking is easy to do.

7. When finished remove both, take the salmon (it can go on a bun opened faced or not) garnish with the cabbage and a couple spoonfuls of your sauce. Put some of your asparagus on your plate and the best salmon you have ever had is ready to enjoy.

ALBUQUERQUE FIRE DEPARTMENT
Firefighter Adrian Breen, Albuquerque, New Mexico

I brag about this recipe because there are a lot of individuals that do not like salmon but I have won everyone I have come across with this recipe. I wish I had a station I am currently in paramedic school. Before medic school. I did 3½ years at station 12. Top 3 out of 23 stations in our department.

E-3
FOR EMERGENCY CALL
911
ENGINE
3
HONORING AMERICA'S
BRAVEST
Pierce

3RD DEGREE BLOCK

It is generally believed that the growing village that was to become Albuquerque was named by the provincial governor Don Francisco Cuervo y Valdes in honor of Don Francisco Fernández de la Cueva y Enríquez de Cabrera, viceroy of New Spain from 1653 to 1660. One of de la Cueva's aristocratic titles was Duke of Alburquerque, referring to the Spanish town of Alburquerque.

Waffle:
4 C. flour
½ C. sugar
2 Tbsp. baking powder
4 eggs
4 C. milk
½ C. oil

Remaining ingredients:
2 packs thick cut maple bacon
1 pkg. jimmy dean breakfast sausage
1 container of maple syrup
1 container of butter
12 eggs
sliced cheddar cheese

1. Prepare the waffle mix and set to the side.

2. Next cook your sausage in a pot, and your bacon on cookie sheets in the oven at 450. Watch the bacon closely it will burn quickly.

3. When your sausage is done set to the side.

4. Get your waffle iron ready.

5. Pour your batter in the waffle iron, along with the two spoonfuls of sausage; infuse the sausage into the waffle.

6. Cook the eggs and all the waffles.

7. When the waffles are done get your plates ready. Place one waffle on the plate, then take 2 eggs and a slice of cheese, and place them on the waffle.

8. Next take 2-3 strips of bacon and place over the eggs.

9. Top all of that with another waffle.

10. Butter your top waffle and pour maple syrup all over. Enjoy the best breakfast sandwich you have ever had.

Firefighter Adrian Breen, Albuquerque, New Mexico

"Good judgment is based on experience, and a lot of that comes from bad judgment."
Will Rogers

Stuffed Zucchini

The Buffalo Fire Department is the largest fire department in Upstate New York. The Fire Department consists of one division which is separated into four battalions, and further separated into four platoons. Each platoon works two day shifts which are from 0800-1700 hrs followed by two night shifts which are from 1700-0800 hrs. The department consists of approximately 675 uniformed firefighters protecting approximately 42 square miles.

zucchini
olive oil
salt
pepper
ground beef/sausage
Spanish spice sofrito
Frank's red hot sauce
red pepper flakes
minced garlic
Philly cream cheese
pecorino Romano
parmesan

1. Halve the zucchini lengthwise.

2. Spoon out contents into a pot on the stove.

3. Paint the remaining skins with olive oil, salt and pepper.

4. Wrap the zucchini skins in saran wrap and toss them in the fridge so they don't brown.

5. Brown some lean ground beef along with any kind of sausage (I like Bob Evans zesty hot breakfast sausage). Use lean ground beef so you barely have to drain it.

6. Add to the browning ground beef/sausage an eyeball amount of the Spanish spice sofrito, some Franks red hot sauce, some red pepper flake, little salt, pepper and a decent amount of minced garlic.

7. Once all has congealed, add it to the pot with the zuch guts, contribute liberal amounts of Philly cream cheese pecorono Romano and maybe some parmesan. Allow it all to come together nicely.

8. After that is complete refrigerate for about an hour so it isn't so soupy.

9. Place cold contents back in skins and bake for one-half hour, 400 degrees L. Enjoy.

"Above all, an assignment to a truck company should be considered a promotion."
John W. Mittendorf

Blueberry Buckle

The Clay Fire Department covers 20 square miles. We currently operate out of two stations with one ladder truck, one heavy rescue, three engines, two EMS squads, a snowmobile and a boat.

1 C. Crisco
¾ C. sugar (creamed together)
1 egg
2 C. flour
2 tsp. baking powder
½ tsp. salt
½ C. milk (add alternating flour mix)
2 C. fresh blueberries

Crumb toppings:
½ C. flour
½ C. sugar
¼ C. salted butter
½ tsp. cinnamon

1. Mix together ingredients; put in greased 9 × 9 pan.

2. Spoon on crumb topping.

3. Bake 35-40 minutes at 375°.

Clay Fire Department
Lieutenant Lou Szitar, Clay, New York

I have been a volunteer at Clay for 35 years and go by the nick name "PAPA-LOU". I was a Lt. and the department treasurer but now I'm just a firefighter.

BUFFALO CHICKEN

The Town of Clay was established in 1827 and is the northernmost town of the nineteen towns in Onondaga County. The largest town in Onondaga County, Clay is 54.6 square miles in size and includes part of the Village of North Syracuse. The Seneca River forms its western boundary, meeting with the Oswego and Oneida rivers at a point known as Three Rivers. The Oneida River forms most of the northern boundary.

6 lbs. boneless skinless chicken
12 oz. Frank & Theresa's anchor bar wing sauce (mild, medium, or suicide)
1 stick salted butter
2 packets of instant chicken bullion
chunky blue cheese dressing
salt & pepper to taste

1. Cut the chicken breasts into quarters and put in crock pot for 6 hours on medium. Enough juice comes out of the chicken that no additional is needed. Break apart the chicken with a fork or potato masher (shredded, like pulled pork). Don't overwork as the chicken can get really fine. Add the Anchor Bar wing sauce, bullion and butter. Salt & pepper to taste. Stir until blended.

2. Serve on a good hard roll or Ciabatta.

3. Can add provolone and/or blue cheese to sandwich.

4. Dress the dish with celery sticks. Enjoy.

When a man becomes a fireman his greatest act of bravery has been accomplished. What he does after that is all in the line of work.
Edward F. Croker

Cornell Chicken Barbeque Sauce

The Clay Volunteer Fire Department was organized on February 19, 1917 as a Fire Protection District. We are 60 of your neighbors, friends and colleagues. We are men and women - married or single - we have full time jobs or go to school - but most importantly we work together to help protect our community. The Clay Fire Department covers 20 square miles.

2 C. cider vinegar
1 C. vegetable oil
1 large egg
3 Tbsp. sea salt
1 Tbsp. Bell's Poultry seasoning
10 lbs. chicken parts with bone in and skin
½ tea ground black pepper

1. Combine the basting sauce ingredients in a blender and blend until emulsified. Place the chicken halves in a large zip-top plastic bag and pour in ½ cup of the sauce. Seal the bag and shake gently to coat the chicken evenly.

2. Refrigerate for 2 hours or more.

3. Grill slowly over coals basting with reserve sauce frequently. Enjoy.

CLAY FIRE DEPARTMENT
Lieutenant Lou Szitar, Clay, New York

If you are dumber than a termite you can be a firefighter. Throw a termite infested log on a campfire...all the termites run out... A firefighter runs in.

SHAMROCK EXPRESS
CHARLOTTE
FIRE
DEPARTMENT

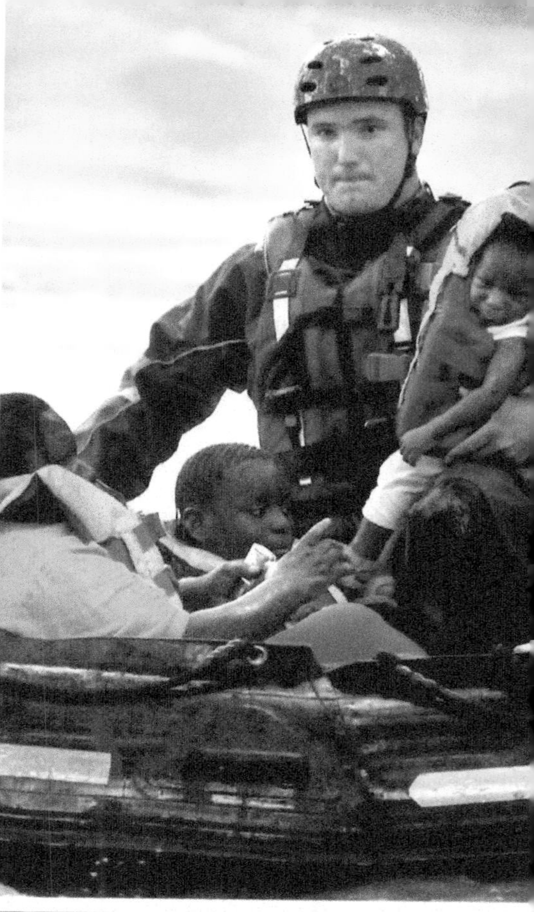

Chicken and Arugula

The Charlotte Fire Department began as robust volunteer firefighters in the 1800s. In the summer of 1887, the City's Board of Alderman voted to hire a full-time Fire Marshal. Today the Charlotte Fire Department approaches its 125th year of service to the community. The Charlotte Fire Department is operating out of 42 fire stations, including 2 stations at Charlotte-Douglas International Airport. Their Fire Chief, Jon B. Hannan, oversees a total of 1,164 personnel, of that 1,044 being firefighters.

1 lb. sliced boneless skinless chicken breast
4 thin slices pancetta, coarsely chopped
2 shallots (2 oz), peeled and thinly sliced
2 cloves of garlic, minced
¼ C. Kerry gold pure Irish butter
⅓ C. drained and chopped sundried tomatoes
½ - ¾ lb. orecchiette pasta, means little ears, any small pasta is fine
salt & pepper to taste
3 C. cleaned and rough chop arugula
1½ C. shaved or grated Dubliner cheese
2 Tbsp. roasted pine nuts

1. Saute chicken, pancetta, shallots, garlic, over heat for about 10 minutes, or until chicken and pancetta are done.

2. Remove from skillet and keep warm.

3. Melt butter in same skillet for 1 minute, or until it turns golden brown, do not burn.

4. Add chicken mixture, tomatoes and cooked pasta to skillet cook to heat through. Season with salt & pepper.

5. Stir in arugula and pine nuts and toss lightly.

6. The most important part - enjoy!

"It's not the depth of the water that kills you, it's the water. More relevant, it's not the size of the fire that gets you, it's the fire."
Jim Smith

Pork Loin Stuffed with Cremini Mushrooms, Basil and Shallots

Also known as the Dilworth fire station, the oldest extant engine house in Charlotte opened in early 1909 with three men, two horses, a combination chemical and hose wagon, and a steam engine. At that time, two years had passed since Charlotte's last volunteer fire company disbanded. Though fully paid firemen had been answering calls since 1887, the department retained on-call personnel. The Neptune Fire Company ceased operation in 1907.

1 boneless pork loin (3-5 lbs.)
2 (8 oz.) packs cremini mushrooms
1 bunch or pre-pack fresh basil
1 medium shallot
4 C. chicken stock/broth
1 Tbsp. Italian seasoning
1 Tbsp. dried basil
1 Tbsp. kosher salt
1½ Tbsp. extra virgin olive oil
2 Tbsp. butter
3 ft. butchers' string
3 Tbsp. corn starch
cracked black pepper to taste

1. Pre-heat oven to 350 degrees.
2. Remove silver skin from loin and butterfly. Set aside.
3. Chop 1 pack mushrooms to your liking. Mince shallot and clean your basil.
4. In non-stick pan sauté mushrooms in 1 tablespoon butter and ½ tablespoon olive oil, when mushroom has a nice caramelized color add shallots and cook until glassine (clear). Set aside.
5. Open loin up, put evenly mushrooms and shallots from one end to the other, place basil leaves on top of that in a single layer.
6. Season inside with cracked black pepper and ½ tablespoon kosher salt.
7. Roll loin into original shape and tie off with butchers' string.
8. Season the outside with rest of Italian seasoning, black pepper, kosher salt and dried basil.
9. Get the same sauté pan hot; add 1 tablespoon butter and ½ tablespoon olive oil. When hot, place loin in and sear all sides dark and colorful.
10. Remove from pan, and put in oven safe dish at least and 1½ to 3 inches deep.
11. Take last pack of the chopped mushrooms and sauté them for a few minutes to get color, add chicken stock to sauté pan and deglaze it, bring to a boil, put in pan with loin. Add 2 tablespoons butter and 5-6 basil leaves to pan.
12. Cover with foil and bake for about 45 minutes. Or until the temperature inside the loin reaches 160 degrees.
13. Take meat and foil and put on cutting board, keep covered. Take the drippings into the sauté pan and bring to a boil. Reduce to about half. Take 2 tablespoons corn starch and 1 tablespoon chicken stock and mix.
14. Stir into boiling liquid and it should thicken, sauce has to be boiling for roux to work. If it is not thick enough repeat this step until your liking. Season with cracked black pepper.
15. Cut off butcher string, and slice. Spoon over sauce and you are ready to go.

This was one of my Captain Sid Cunningham's favorites; he still talks about it to this day.

Charlotte Fire Department
Firefighter Jeff Nixon, Charlotte, North Carolina

Station 39's Carolina BBQ

The Charlotte Fire Department operates out of 41 Fire Stations and provides fire protection for approximately 300 square miles. (Station 42 under construction) With 1,164 full-time positions, 1,044 are fire suppression personnel, 41 Engine Companies and 15 Ladder Companies, Charlotte Fire Department responded to over 93,000 calls for service in 2010.

6-8 lb. Boston butt (trimmed)
1 Tbsp. olive oil
2 Tbsp. cracked black pepper or to taste
2 Tbsp. Red Monkey tres chili and cilantro powder
2 Tbsp. Red Monkey smoked paprika and roasted garlic
1 Tbsp. Captain Wetta's Magic Dust (2 Tbsp. of regular chili powder is acceptable substitute)
2 Tbsp. garlic powder
1 Tbsp. Italian seasoning
1-1½ qts. chicken stock (enough to cover ¾ of butt in crock pot)
1 (18 oz.) bottle of Budweiser Smokey BBQ sauce
¾ bottle KC Masterpiece Hickory BBQ sauce
1 Tbsp. Gravy Master

1. Turn on grill to high.

2. Trim and rub olive oil on Boston butt. Cover all sides of pork with seasonings and rub in.

3. Put butt on grill for approximately 3-5 minutes (depending on size) and then repeat on all sides, until well seared and has achieved nice caramelization.

4. Put butt in crock pot set on high, and add BBQ sauces and chicken stock.

5. Add 1 tablespoon Gravy Master, cover and cook for 3½ -4½ hours until internal temperature reaches 165° F, or meat falls off the bone.

6. Remove pork, and transfer to cutting board. Pull apart or chop, depending upon your preference. Add pork to the choice of sauces - Eastern Carolina (vinegar based) or Western Carolina (traditional sauce)

Eastern: Take 6 ounces ladle of drippings from the crock pot and mix with pork. Add 8 ounces of Scotts BBQ sauce - (it's vinegar based, so it's thin.) Add 3 tablespoon Sweet Baby Ray's BBQ sauce, and serve on a bun with coleslaw.

Western: Take 6 ounces ladle of drippings from crock pot, and mix with pork. Add 10 ounces Sweet Baby Ray's BBQ sauce, and serve on a bun with coleslaw.

Firefighter Jeff Nixon, Charlotte, North Carolina

"You have to do something in your life that is honorable and not cowardly if you are to live in peace with yourself, and for the firefighter it is fire."
Larry Brown

Chili Con Queso

The Dickinson Volunteer Fire Department is responsible for public education concerning fire and life safety issues, enforcement of the Fire Code, the recruitment and training of volunteer firefighters, fire suppression and hazardous materials incident response, and fire investigation for cause and origin. Full time staff is also responsible for the procurement and maintenance of all necessary equipment.

1 lb. Jimmy Dean spicy breakfast sausage
2 lbs. hamburger
2 lbs. block of Velveeta cheese (not Velveeta Light)
2 medium to large tomatoes
1 green bell pepper
1 red bell pepper
2 medium size onions (1 white & 1 purple)
1 or 2 bags of Frito Scoops

Optional:
1 Jalapeno or Tabasco sauce

1. Brown the hamburger and sausage in a skillet.

2. While cooking that up, cut up the cheese block into chunks and place in large Crock Pot.

3. Dice up all of the other vegetables and give them a toss into the Crock Pot with the cheese.

4. Once the hamburger and sausage are done, drain the meat and place into Crock Pot with rest of ingredients.

5. Turn on high and begin the melting process and be sure to give it an occasional stir.

6. Once it is all melted down, lower cooking temperature and continue cooking for about 1-2 hours to soften the veggies. When done, grab yourself a cereal bowl and scoop in the Con Queso (add Tabasco sauce to taste), have another bowl for the Frito Scoops and begin dipping and enjoy. Great for watching a football game.

Dickinson Volunteer Fire Department
Captain Kent Mortenson, Dickinson, North Dakota

The first large organized force of firefighters was the Corps of Vigiles, established in ancient Rome in 6 AD.

Stuffed Biscuits/ Crescent Rolls

The goal of the department is to prevent fires and other emergency incidents by maintaining an aggressive public education and fire code inspection program. To have a well-trained, well-equipped response team available to mitigate incidents that do happen. To insure that all equipment is kept in a ready state, whether it is a fire engine or firefighters' personal protective equipment.

ham
cheese
roast beef
diced onion, optional
green peppers, optional
Pillsbury butter milk biscuit or crescent rolls

1. Determine was you want as a stuffing: ham and cheese, roast beef and cheese (diced onion or green peppers optional).

2. Either cook you own meat or pick up some sliced meat at your local deli and buy a couple packs of whatever type of shredded cheese you favor.

3. Then find yourself a couple of the cup cake type pans that usually make about 1 dozen cupcakes.

4. Then use either your family secret biscuit/crescent roll recipe or pick up some Pillsbury Butter Milk Biscuit or Crescent Rolls at the grocery store.

5. Lightly spray your cupcake tins, open up or prepare your dough nice and flat to allow you to stuff into the cupcake tin.

6. Once you have the dough in the tin, place you're filling of meat (beef, ham, pork or chicken) in the tin and cover the shredded cheese.

7. Then take your corners of the biscuit/crescent dough and fold it up and bake. Just following regular baking instructions as on the container, as your location altitude may vary.

Note:
If you like a little spice, try dicing up some chicken breast and cook it up in a pan with some Cajun spice for your filling ingredient. If you like the biscuits make them in buttermilk biscuits. Don't be afraid to think outside the box and use some pizza sauce and pepperoni.

Here's another quick and simple way to feed a lot of people with ease! Try making and serving stuffed biscuits or crescent rolls. Everyone loves Pillsbury biscuits or crescent rolls, especially when they are hot out of the oven.

Dickinson Volunteer Fire Department
Captain Kent Mortenson , Dickinson, North Dakota

Chicken Spinach Penne Pasta

Welcome to the City of Green, Ohio. Located in Northeast Ohio between Akron and Canton, our city is a progressive, prosperous and promising city for residents and businesses alike. In recent years, Green was named one of the best places to raise a family by Businessweek.com.

3 lbs. of diced chicken breast
2 (14.5 oz.) cans of diced tomatoes with basil/garlic and olive oil
1 jar of pesto 8-10 oz. size
1 small box of frozen chopped spinach
1 lb. of penne pasta
2 C. shredded Italian cheese

1. Use a large fry pan on medium/medium-high heat. Oil pan with olive oil and cook diced chicken breast. Drain liquid and then add all ingredients above except pasta and shredded cheese. Bring to a simmer and cook for 10-15 minutes.

2. Boil pasta until tender and then toss all together in a large serving bowl with shredded Italian cheese.

3. Served with salad and garlic bread.

"...the man who really counts in the world is the doer, not the mere critic-the man who actually does the work, even if roughly and imperfectly, not the man who only talks or writes about how it ought to be done."
Theodore Roosevelt

Breakfast Rollups (Taquitos)

Built on former reservation lands of Kiowa, Comanche, and Apache Indians, Lawton was founded on August 6, 1901, and was named after Major General Henry Ware Lawton, a Civil War Medal of Honor recipient who was killed in action in the Philippine–American War. Lawton's landscape is typical of the Great Plains with flat topography and gently rolling hills, while the area north of the city is marked by the Wichita Mountains.

1 pkg. flour tortillas, small ones
1 doz. eggs
1 lb. pork sausage (J.C. Potter-hot is best 'cause it's made in Oklahoma ya'll)
2 med. size potatoes very finely diced (frozen diced potatoes or hash browns may be substituted)
1 (8 oz.) pkg. of Kraft Mexican
4 cheese or Colby/Monterey Jack
1 jar of your favorite picante Sauce
½ C. milk or cream
salt and pepper to taste

1. Brown diced potatoes or hash browns, salt and pepper to taste. Brown sausage, breaking it up into very small pieces.

2. Drain grease from potatoes and sausage.

3. Place them in a bowl with lots of paper towels under them to absorb excess grease.

4. Scramble eggs with with ½ cup milk and salt and pepper to taste. Scramble eggs in skillet large enough to hold all other ingredients. Use butter or margarine (¼ stick or less) to cook eggs.

5. When eggs are about ¾ solid, pour in sausage and potatoes.

6. Fold the mixture together and finish cooking until eggs are firm.

7. Cover and keep warm until served.

8. Place tortillas on microwave safe plate.

9. Cover with inverted plate of same size.

10. Microwave for 1 to 1½ minutes.

11. Put egg, sausage, potato mixture on tortilla. (not too much or you can't roll it up)

12. Sprinkle cheese on filling. Add Picante to taste. Roll 'em up. (moisten the edge with Picante to seal it like an envelope) Yo Quiero Firehouse Rollup!

Lawton Fire Department
Firefighter Lt. Randy Britton, Lawton, Oklahoma

In 1648, Governor Peter Stuyvesant of New Amsterdam (New York City) was the first in the New World to appoint fire inspectors with the authority to impose fines for fire code violations.

7UP Salmon

The Oklahoma City Fire Department was once a small volunteer fire company with one donated, hand-drawn wagon housed in a small-framed building. When formed it was the Department's duty to protect less than 10,000 residents. Today, the responsibility has grown to include over 600,000 residents. The Department has grown to approximately 1,000 career personnel operating around the clock. Our Department has 35 strategically located fire stations that respond to over 88,000 emergencies per year

4 salmon steaks (4-6 oz. each)
3 onions, peeled and chopped
3 tsp. instant chicken bullion
1 tsp. ground mustard
1 tsp. cumin
1 tsp. ginger
2 tsp. salt
½ C. water
1 C. 7UP (or Sprite)

7UP Sauce:
3 Tbsp. flour
1 can condense
2 tsp. salt
1 Tbsp. lemon juice
3 Tbsp. butter or margarine
1 C. 7UP (or Sprite)

1. Arrange Salmon steaks in an 8″ x 8″ glass baking dish. Top with onion, bullion, seasonings, water, and 7UP. Cover with wax paper. Cook at 450 F until fish flakes easily. Drain liquid. Pour 7UP sauce over top to serve.

7UP sauce:
1. Combine all ingredients, except butter, in a large bowl. Rotary beat until smooth. Add butter. Microwave on high for 4 minutes. Beat lightly.

"A fire is like being pregnant, either you have one or you don't"

Blueberry Dump Cake

Oklahoma City was first settled on April 22, 1889, when the area known as the "unassigned lands" (that is, land in Indian territory that had not been assigned to any tribes) was opened for settlement in an event known as "The Land Run". Some 10,000 homesteaders settled the area now known as Oklahoma City; the population doubled between 1890 and 1900.

1 can of blueberry pie filling (any type of pie filling may be substituted instead of blueberry)
1 can of crushed pineapple
1 butter pecan cake mix
1 stick of real butter

1. Mix pie filling and can of pineapple together without draining in a 9 x 13 cake pan.

2. Cover the mixture with the cake mix.

3. Melt the stick of butter in microwave and pour evenly over cake mix.

4. Pre-heat oven to 375 and bake for about 45 minutes.

5. Type of pan will determine baking time.

6. It is done when top is brown.

Oklahoma City Fire Department
Firefighter David Mitchell, Oklahoma City, Oklahoma

"I think that anyone who says that he can predict a fire season with any accuracy has to be either absolutely brilliant, or God's own fool and I'm not either one. So I'm not even gonna try."
CalFire Supervisor

C F D

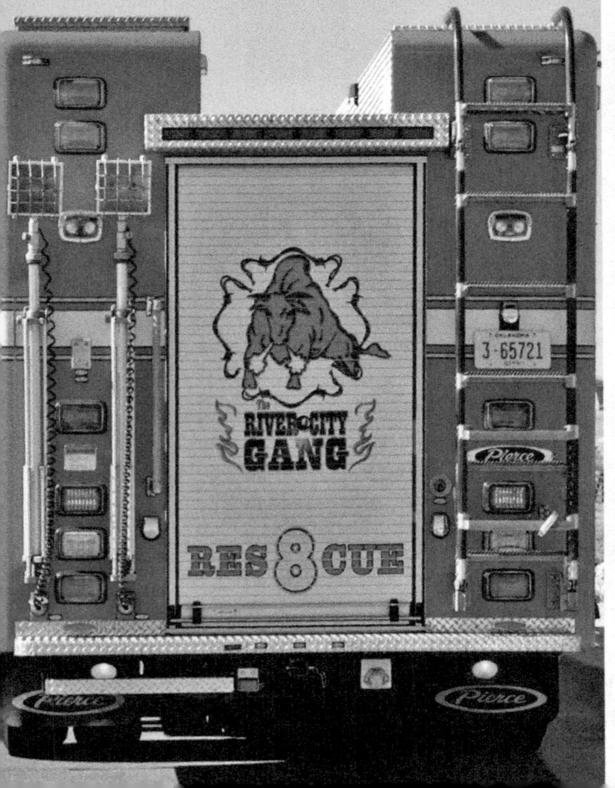
The
RIVER CITY
GANG
RES 8 CUE
3-65721
Pierce
Pierce
Pierce

OKLA CITY
CORPORAL
E17
SCOTT

E54

PARAMEDIC
ENGINE 51
FIRE DEPT.
51
OKLAHOMA CITY
FIRE DEPT.

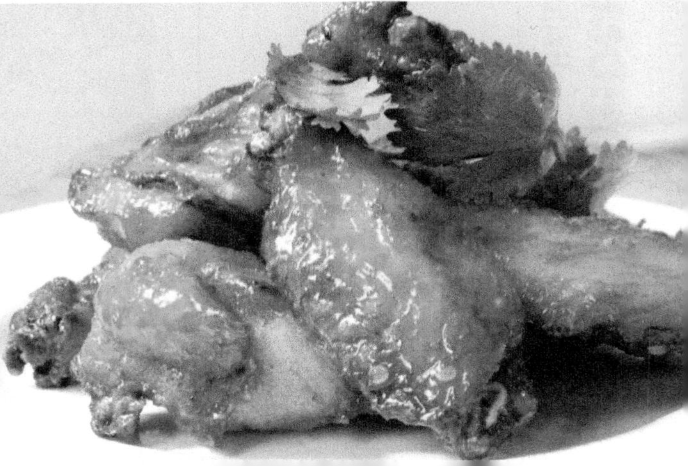

CAJUN CHICKEN PASTA

Firefighters Brian Suchy and Stephen Lewis, Oklahoma City, Oklahoma

Nineteen eighty-nine was a pivotal year for the Oklahoma City Fire Department. In retrospect, one can see that the events of that year prepared the department for what would be it's greatest challenge thus far, the rescue and recovery at the Oklahoma City bombing.

1 lb. chicken breast
1 (16 oz.) pkg. bowtie pasta
1 green bell pepper
1 red bell pepper
1 medium white onion
1 pkg. sliced mushrooms
1 large tomato
1 cucumber
cajun spice
salt and pepper
olive oil

1. Cut chicken into 1 inch strips and sauté using Cajun seasoning to taste.

2. Bring water to a boil and cook bowtie pasta until it is al denté.

3. Chop peppers, onions, and mushrooms and sauté in a separate pan with 2 tablespoons of butter, until soft.

4. Once pasta is done, drizzle in olive oil and add Cajun seasoning to taste.

5. Add the cooked chicken, pepper mix and pasta into a large serving dish and mix well.

6. The cucumber is to be sliced and placed over the hot pasta mix, dice the tomato and garnish the dish.

Boston imported (1679) the first fire engine to reach America. For a long time the ten-person pump devised by the English inventor Richard Newsham in 1725 was the most widely used.

Chicken and Dumplings

Nineteen eighty-nine was also the department's centennial. In 1889, the department began as a volunteer bucket brigade. Its first piece of apparatus was an old beer wagon equipped only with ladders and buckets, pulled by hand to fires.

1 whole chicken
2 C. of flour
1 tsp. baking powder
1 egg
4 Tbsp. shortening
4-6 Tbsp. milk
1-2 cans chicken broth, (may or may not need)
pepper
garlic powder

1. Boil whole chicken till done, keep broth and cook it down until it is rich enough, you can add the bones and skin from the chicken until it cooks down, strain broth and set aside. De-bone the chicken and chop up, add some garlic powder, be sure to pay attention and get the gristle out.

2. To make the dumplings, mix the dry ingredients, mix the liquids together and add to the dry stuff and mix, divide dough in two parts and roll out thin on a lightly floured board, cut with a sharp knife ¼ to ½ wide, drop one at a time into boiling broth, cover and simmer 12-15 minutes.

3. You can add 1 to 1½ cups of water if broth is too rich, or butter if it is not rich enough. Use canned broth as needed.

4. Salt and pepper as desired.

Firefighters Brian Suchy and Stephen Lewis, Oklahoma City, Oklahoma

A man may build a complicated piece of mechanism, or pilot a steamboat, but not more than five out of ten know how the apple got into the dumpling.
Edward A. Boyden

DUMP CAKE OR QUICK COBBLER

By 1891 finances had improved to the point that two horses, Babe and Jumbo, were purchased to do the pulling. Firemen were justly proud of their horses, who reportedly could distinguish the ring of the fire phone from the local phone and were always in place before the firemen had time to slide down their poles.

1 yellow cake mix
1 stick of butter (real butter, margarine will not work), softened
2 regular cans of fruit pie filling

1. Heat oven to 350 degrees.

2. Pour fruit filling in a cake pan, spreading evenly.

3. Put cake mix in a large bowl and add softened butter.

4. Hand or spoon mix until cake mix is moist.

5. Mixture should be crumbly.

6. Spoon the mixture onto the fruit filling evenly until the filling is covered.

7. Bake uncovered at 350 degrees for 30 - 35 minutes.

The very quick and less messy way:
1. You do not need to have the butter softened.

2. Spread fruit filling evenly in cake pan.

3. Pour cake mix on top of filling, spreading evenly.

4. Slice and put thin pats of butter evenly spread on top of cake mix.

5. Bake uncovered at 350 degrees for 30 – 35 minutes.

OKLAHOMA CITY FIRE DEPARTMENT
Firefighter Steve Baggs, Oklahoma City, Oklahoma

"Tactical catastrophes are never the outcome of a single poor decision. Small compromises incrementally close off options until a commander is forced into actions he would never choose freely."
Nathaniel Flick

Firehouse Chicken Wings

No matter how much they denied it, they were our heroes, symbols of sanity amidst a landscape of madness. And of course we were right, too, if you define a hero as someone noted for feats of courage or nobility, as my dictionary does.

4-6 lbs. of chicken wings
one bottle of Franks hot sauce
½ stick of butter
4-5 Tbsp. olive oil
cracked black pepper
sea salt
garlic powder

1. Preheat oven to 425 degrees.

2. Take wings and cut to separate drum from wing section.

3. Take a large non-stick cookie sheet and spray with non-stick spray.

4. Coat pan with olive oil.

5. Place chicken wing pieces on cookie sheet.

6. Sprinkle with cracked black pepper, sea salt, and garlic powder.

7. Place in oven for one hour. In a small bowl put together Franks hot sauce and butter.

8. Place in microwave for 30 seconds till butter melts. Mix well.

9. After one hour turn wings, cook for 15 more minutes.

10. Remove wings from oven.

11. Use tongs to dip wings one at a time into the sauce, roll until coated and place on a serving sheet.

Oklahoma City Fire Department
Firefighter Dante Viviani, Oklahoma City, Oklahoma

GREAT BRISKET

Agreeing with the public, the International Association of Fire Chiefs awarded the Oklahoma City Fire Department the Ben Franklin award for heroism and valor. Usually given to an individual, the medal is the highest honor the fire service bestows.

1 Saddlers brand, already sliced brisket from Sam's Club
4 packs of Lipton onion soup mix
2 large onions
2 packs of whole uncooked mushrooms

1. Place meat into a large high sided pan. You may use aluminum pans but double the pans because it is heavy when handling the pans.

2. Lining with foil will also add to the strength of pans.

3. Mix Lipton soup packets with about ½ gallon of water and pour mixture over brisket that has been placed into pan.

4. Cover the top of the pan with foil and place into oven, pre-heated to 200 degrees.

5. After cooking for 2 hours take out of the oven and pour the juice into a container that can be used to easily pour back onto the brisket.

6. Pry the pre-sliced brisket apart as much as you can and pour the mixture back on it and place back into the oven for 2 more hours.

7. After 4 hours put the sliced onions and mushrooms into the pan and do this again every 2 hours for a total of 8 hours cooking time.

8. If onions and mushrooms do not get done you may turn the heat up on the oven and cook for a little while longer.

9. These briskets are sliced thin and will feed a large amount of people.

OKLAHOMA CITY FIRE DEPARTMENT
Firefighter David Mitchell, Oklahoma City, Oklahoma

The inventor Thomas Lote of New York built (1743) the first fire engine made in America. About 1672 leather hose and couplings for joining lengths together were produced; though leather hose had to be sewn like a fine boot.

LOUISIANA POTATO SALAD

"I couldn't be prouder," says Fire Chief Gary Marrs of his firefighters. In a letter of commendation each firefighter received in June, Marr wrote: 'You have shown this community and this country the best part of human nature: the willingness to help others in time of great need ...You have proved that this is the best fire department in the nation."
By Maura McDermott, Oklahoma Today

5-6 lbs. medium red potatoes (peeled, cut into chunks, and boiled)
16 eggs (hard boiled and cut into chunks)
1 lb.-8 oz. smoked bacon (chopped and fried crispy)
1 C. chopped celery
¼ C. chopped scallions
¼ C. chopped parsley
5 C. mayonnaise
1½ C. Miracle Whip
1 small jar creole mustard
2 tsp. salt
1 tsp. cayenne
1 tsp. black pepper

1. Prepare potatoes, bacon, and eggs, set aside to cool.

2. Thoroughly mix the remaining ingredients.

3. Gently fold in the potatoes, eggs, and bacon.

Ladder trucks provide access to buildings as much as 100 ft (30 m) high; snorkel trucks enable firefighters to douse fires from above.

MOP HEAD STEW

The Oklahoma State Firefighters Museum is a museum owned and administered by the Oklahoma State Firefighters Association (OSFA). The museum is financed by the dues collected from more than 8,000 firefighters, and is located at 2716 N.E. 50th Street in Oklahoma City, Oklahoma.

1 small roast (beef or deer roast)
2 (16 oz.) cans tomato sauce
4 cubes beef bullion
2 cans sliced carrots
1 can green beans
2 cans corn
1 white onion
4 small potatoes (optional)
salt and pepper to taste

1. Sear roast in a skillet using olive or some other type cooking oil.

2. Place roast in a cooking bag or crock-pot and pour in the 2 cans tomato sauce.

3. Cover the roast the rest of the way with water. Add the four cubes of beef bouillon.

4. Cook the roast for at least 4 to 6 hours or until the meat is tender and falling apart.

5. Take the roast out of the pot and place on a cutting board for trimming.

6. While the roast is cooling you can add the vegetables to the broth you have in the pot.

7. If you use potatoes you will have to have them in with the roast from the beginning.

8. Once you have the vegetables in, trim the fat off the roast and cut the meat to the size you prefer.

9. Add this back to the stew, salt and pepper to taste.

10. Cook till all vegetables are tender.

"There is no fire more important than the one that is currently burning."
Pete Pedersen, Fire Chief (Ret)
Speaking to a political body, defending his decision to send a large amount of resources to a mutual aid incident.

Wolf's Skillettini

Groundbreaking for the museum building was performed on April 6, 1967 and was completed in 1969. The building houses the museum, the (OSFA), the Oklahoma Fire Chiefs Association (OFCA), the Council on Firefighter Training (COFT) and the Oklahoma Retired Firefighters Association (ORFA) offices.

3 boneless skinless chicken breast
1 pkg. thin spaghetti noodles
1 pkg. parmesan cheese
5 Italian sausage links
1 bell pepper
1 onion
1 jar red spaghetti sauce
2 cloves garlic

1. Grill chicken and sausage.

2. Put your noodles on to cook.

3. Dice your onion and bell pepper into 1 inch squares and set aside.

4. When meat is done set to the side and let cool, then cut up into slices.

5. When noodles are done rinse with cool water and let set. Put your sauce on the stove to warm up.

6. In a large cast iron skillet sauté your bell pepper, garlic and onions in olive oil.

7. When onions and peppers are soft add your meat.

8. When meat is added, put your noodles in a handful at a time.

9. As adding noodles roll them around in the skillet until heated through.

10. Then add red sauce a little at a time, when heated through top with the parmesan cheese and serve.

Oklahoma City Fire Department
Fire Chef Roy Wolf, Oklahoma City, Oklahoma

Oil-field fires demand multiple approaches: water streams, fogs, foams, and explosives may all be used simultaneously to quench a fire and prevent its reignition.

DRAGON'S BREATH BOLOGNA

The Tulsa Fire Department was organized as a volunteer department on June 6, 1900 and became a four-man paid department on December 6, 1905, two years before statehood. In 1913 Tulsa became the first completely motorized department west of the Mississippi River

5 lbs. roll of bologna
dirty rice, for stuffing
1 pkg. hot links, chopped finely
Chipotle/Wasabi BBQ sauce

Chipotle/Wasabi BBQ Sauce:
1 C. BBQ
3 Tbsp. lemon juice
¼ C. canola oil
1 tsp. Dijon Mustard (1 Tbsp. if omitting Wasabi)
1 tsp. wasabi powder
2 Tbsp. chipotle peppers in adobo sauce
¼ tsp. red pepper flake
¼ tsp. cayenne
ground pepper to taste

1. Cook dirty rice according to directions adding the chopped hot links in the mix. Core out the center of the bologna roll with a thin walled galvanized pipe (preferably a clean one) and use 2 inches of the center as plugs.

2. Stuff the dirty rice mixture into the center of bologna, plug and toothpick the ends to hold it all.

3. Wrap in foil and smoke for 2-4 hours.

4. Serve with sides of your choice & chipotle sauce, large glass of something cool to drink & pepto in the "staging area" for all the light weights.

Chipotle/Wasabi BBQ Sauce:
1. Mix in blender thoroughly. Serve along side dish and pre-baste dish before smoking… if you have a good "water supply" established!

The inventor Ctesibius of Alexandria devised the first known fire pump c.200 but the idea was lost until the fire pump was reinvented about 1500.

Toro Diablo

The Tulsa Fire Department has an authorized strength of 724 employees, 694 uniformed and 26 civilian personnel. The Department has 30 Fire Stations housing 30 Engines, 12 Ladder Trucks, a HazMat Unit, provides Technical Rescue and Airport Firefighting capabilities. The operating budget for 1995 was $38.6 million.

3-4 lbs. beef arm roast or boneless pot roast
3 baker potatoes, peeled & sliced thinly
1 onion, sliced
1 Tbsp. Dijon mustard
1 Tbsp. sriracha
1 tsp. vinegar
2 Tbsp. flour
1 Tbsp. Worcestershire sauce
1 tsp. sugar
1 bag small carrots

1. Trim excess fat from roast and place in crock pot on top of onions and potatoes.

2. Make a paste with the remaining ingredients and smear onto the roast.

3. Cover and cook on "smoldering" aka Low 10-12 hours (or on "Getting' It" for 5-6 hours). Ring the dinner bell and get to it before the drop hits!

Tulsa's paid department began with a continuous duty system and implemented a twenty-four on twenty-four off (72-hour workweek) in 1919.

Kraken Stew

In the early 1970's recognizing the importance of improved firefighter safety and health standards the Department began purchasing five man cab apparatus to get firefighters off the tailboard. In 1978 personnel safety policies required firefighters utilize (ride in) the apparatus cab rear seats, wear full protective clothing, and mandated the use of self contained breathing apparatus.

2 lbs. shrimp, shelled & cleaned
1 Zatarans shrimp boil spice bag
1 pkg. crab meat
3 Tbsp. tomato paste
1 can cream of shrimp soup
2 Tbsp. water
2 tsp. dry sherry
3 Tbsp. milk
2 Tbsp. flour
rice or pasta for 4 firefighters
Bread for "Salvage" of leftovers
salt, pepper & garlic salt to taste

1. Place all ingredients (save the milk, flour & rice/pasta) into a crock pot & cook on "Getting' It"aka High for 3-5 hours.

2. Remove spice bag. Just before calling in the crew, make a paste of the milk and flour, adds to the pot and serves over the foundation of choice.

3. Salvage is a must so clean those plates well!

Salvage operations consist of those methods and operating procedures allied to fire fighting that aid in reducing fire, water, and smoke damage during and after fires.

Loaded Baked Potato Soup

After the American Civil War, numerous settlers, including Joshua Welch, George Betts, Charles Angel, W. P. Watson, and John Henry, laid out what is now known as Beaverton hoping they could bring a railroad to an area once described as "mostly swamps & marshes connected by beaver dams to create what looked like a huge lake."

1 bag of frozen shredded hash browns 30 oz. size
1 can cream of chicken soup
3 cans of chicken broth
½ C. minced onion
pepper to taste
1 block of cream cheese (fat free won't melt)

1. Combine all ingredients in a crock pot except for the cream cheese and cook on low all day. 1 hour before serving, you add in the cream cheese and stir occasionally.

2. And then you can top with bacon, cheese or sour cream.

3. You can add other things to the soup like chopped ham or kielbasa sausage. Definitely a crowd pleaser! And it is super easy.

Beaverton Fire Department
Facilities Assistant Samantha M. Smith, Beaverton, Oregon

The Beaverton Fire dept. was first organized in the early 1970's with only a few members and they were James Pickle, Cloves Cannon, Guy White and Frankie Guyton.

Russian-Ukrainian-American Pancakes

The name of the city is taken from the Tualatin River, which flows along most of the city's northern boundary. It is probably a native American word meaning "lazy" or "sluggish" but possibly meaning "treeless plain" for the plain near the river or "forked" for its many tributaries. According to Oregon Geographic Names, a post office with the spelling "Tualitin" was established November 5, 1869, and the spelling changed to "Tualatin" in 1915.

2 C. of cottage cheese
¼ C. of sugar
1 C. of cream cheese
3 eggs
1 Tbsp. Costco vanilla extract
¼ of tsp. of baking powder
1 C. of flour

1. Blend cottage cheese and cream cheese together.

2. Move the paste to the bowl and add vanilla, eggs and sugar. Mix it all together.

3. Mix in flour and baking powder.

4. Preheat the skillet on low-medium (or 350 degrees); put some vegetable oil on it.

5. Put the dough in 4 inches circles on the skillet and flip them when they turn golden color. Enjoy!

The Tualatin Valley and Beaverton area was originally the home of native people who referred to themselves as Atfalati.

TACO SOUP

Tualatin Valley Fire & Rescue provides fire protection and emergency medical services to approximately 440,000 citizens in one of the fastest growing regions in Oregon. Our 210 square mile service area includes nine cities and unincorporated portions of Clackamas, Multnomah, and Washington County.

1.5 lbs. ground beef browned
2 jars of Pace Picante salsa (16 oz.) (medium)
2 cans of corn (drained)
2 cans of kidney beans (drained)
1 large can of crushed tomatoes

1. Brown ground beef.

2. Drain off grease and add the remaining ingredients.

3. Bring to a boil and simmer for 1 hour top with sour cream and cheese and serve with Fritos corn chips or tortilla chips.

BEAVERTON FIRE DEPARTMENT
Facilities Assistant Samantha M. Smith, Beaverton, Oregon

Tualatin Valley Fire & Rescue is an internationally recognized fire department. In 1999 and again in 2006, TVF&R received the prestigious International Association of Fire Chiefs' Fire Service Award for Excellence.

Firefighter Lieutenant Neil Martin, Portland, Oregon

APPLEWOOD SMOKED FREE RANGE CHICKEN

In the year 1850, a company of volunteer firefighters was organized with Thomas J. Dryer, owner of Portland's first newspaper, The Weekly Oregonian, as foreman. Through voluntary subscription the organization, which was known as Pioneer Fire Company No. 1, had acquired a hand pump, which was many times more effective in extinguishing fires than the earlier bucket brigade.

6 free range whole chickens (2.5 to 3 lbs. each)
salt and pepper

BBQ Sauce Ingredients:
3 C. tree top apple juice
4 C. ketchup
¼ C. each red wine, balsamic, and sherry vinegar
1 C. honey
□ C. soy sauce
2 Tbsp. Dijon mustard
□ C. Worcestershire sauce
5 cloves garlic, minced
5 sprigs rosemary
2 Tbsp. olive oil

1. Place Tree Top apple juice in a non-reactive sauce pan over medium heat and reduce by half. Mix all ingredients in a large bowl. Cover and place it in the refrigerator overnight.

2. Season the chicken with salt and pepper.

3. Trusse the chicken with some cotton butcher string.

4. Place the chicken directly over the hot mesquites. Place the lid on the Webber grill and make sure the top vent is open halfway.

5. Let the chicken cook for 45 minutes before removing the lid.

6. At this time pour the BBQ sauce over the chicken. Do not brush!

7. Repeat this three times placing the lid over the chicken each time. When the chicken has reached 145 degrees internal temperature, remove from the heat and let it rest in a warm place for 30 minutes.

"That any person or persons may be called upon to assist the Fire Department and failure to comply with this order is punishable by fine . . .Physicians, whilst engaged in their professional duties, are exempt from this order . . ."
Portland Ordinance No. 11, May 22, 1854

FIRE
RESCU

HOT SAUCE
ORIGINAL

FIRE DEPT

Waldorf Salad

In 1843, William Overton saw great commercial potential for this land but lacked the funds required to file a land claim. He struck a bargain with his partner, Asa Lovejoy of Boston, Massachusetts: for 25¢, Overton would share his claim to the 640-acre site. Overton later sold his half of the claim to Francis W. Pettygrove of Portland, Maine. Pettygrove and Lovejoy each wished to name the new city after his respective home town. In 1845, this controversy was settled with a coin toss, which Pettygrove won in a series of two out of three tosses

3 C. Tree Top apple sauce
2 C. plain yogurt
juice of one lemon
6 each sweet crisp Washington apples
12 each heads of yellow endive
6 stalks celery with leafs
2 C. toasted hood river walnuts
1 C. rouge river blue cheese
ground black pepper

1. Combine the Tree Top apple sauce, yogurt and lemon juice in a large bowl. Use a large wire whisk to emulsify the ingredients. Season with ground black pepper to taste. Place in the refrigerator make sure to cover with clear wrap.
2. Peel and core the apples and place in a bowl of salted cold water for 30 minutes, this will keep the apples a nice crisp color.
3. Slice the endive in half length wise and cut out the core.
4. Julienne the endive into ¼ inch strips and place in a large bowl covered with a damp towel and place in the refrigerator.
5. Remove the leaves from the celery. You should have about a cup of leaves.
6. Slice the celery in half and julienne the half pieces.
7. Remove the apples from the water and dry on a towel.
8. Julienne the apples the same size as the celery. It's very important to keep ingredients very cold until you combine right before you serve.
9. Combine the endive, celery and apples in a large bowl.
10. Pour half the dressing around the bowl and mix gently.
11. Top with the walnuts, blue cheese and celery leaves, you can mix gently if you like, adjust the amount of dressing to your taste and serve.

Portland Fire Department
Firefighter Dustin Miller, Portland, Oregon

The City at present has more than 80 fire cisterns located at convenient points in the metropolitan area, dating from the 1800's.

Pot Roast Ala Jack

Croydon does not have a central government of its own. It is not organized or incorporated as a town or village. The area known as Croydon is located in Bristol Township. Croydon has an all-volunteer fire company which handles fire, medical, marine, and other calls servicing the community named Croydon Fire Company #1.

salt and pepper to taste
1 (3-lbs.) chuck roast (with bone in it)
5 carrots
5 celery stalks
1 large onion
4 Idaho potatoes (or 8 new, or red potatoes)
1 (10¾ oz.) can cream of mushroom soup (Campbell's preferred)
1¼ cans water
1½ Tbsp. Dijon mustard
1 Tbsp. gravy master

1. Salt and pepper roast on both sides.

2. Place roast in good-size pan and cook in pre-heated 400-degree oven for 20 minutes to brown.

3. While meat is cooking, clean and cut carrots into 2-inch pieces. Repeat process with celery stalks.

4. Peel and halve the potatoes.

5. Mix mushroom soup and water in large bowl. Add mustard and Gravy Master for color and flavor.

6. Mix well. Remove roast from oven and place vegetables around meat in pan. Pour sauce mixture over meat and vegetables, cover with aluminum foil, and bake for 2 hours at 350 degrees.

7. Should be served with Dewer's White Label scotch on rocks. (if desired)

Croydon Fire Department
Firefighter Jack Sassani, Croydon, Pennsylvania

Established in 1918, the fire department has been serving the Croydon community for over 90 years.

Blue Crabs, Hot & Spicy

The eponymous King of Prussia Inn was originally constructed as a cottage in 1719 by the Welsh Quakers William and Janet Rees, founders of nearby Reeseville. The cottage was converted to an inn in 1769 and did a steady business in colonial times as it was approximately a day's travel by horse from Philadelphia

2 cans (12 oz.) of your favorite beer. (Labatt's Blue Light works for me)
1½ to 2 C. of white vinegar
2 C. of water
hot sauce, about 6 to 8 shakes
¼ C. of old bay seasoning
1 Tbsp. of crushed hot red pepper
10 blue crabs

1. Turn the heat on high.

2. While the pot is warming, place 1¼ cup of old bay in a bowl, add ¾ cup of kosher or sea salt and mix thoroughly. Keep this aside for now.

3. When the pot comes to a boil, layer the crabs, one layer at a time. On each layer sprinkler the Old Bay and salt mixture, generously to the crabs. Also, shake some (be generous) crushed hot red pepper on the layer as well. Start second layer and repeat until all crabs are in the pot.

4. Let the pot boil for 20 minutes, remove from heat, remove crabs and serve.

"You didn't start the fire, it's not your fault. Just do the best you can to make it better."
Gene Sadler

Ricotta Stuffed Eggplant Rolls

Established in 1854, Providence is the second oldest professional fire department in the country and continues to set the standard for modern professional firefighting and emergency response while being regarded as some of the most aggressive and well trained in the country.

Rhode Island Fire Academy
Director Joe Castro, Providence, Rhode Island

1 eggplant
extra virgin olive oil
1 lb. ricotta cheese
¼ C. pecorino cheese
5 basil leaves, chopped
salt and pepper
2 eggs
½ C. shredded mozzarella cheese

1. Slice the eggplant lengthwise on a mandolin □ of an inch thick.

2. Heat the grill or a grill pan to high heat.

3. Lightly coat each piece of eggplant with oil and place on the grill for 1 minute.

4. Flip the eggplant and allow to cook for one more minute.

5. Combine the ricotta, pecorino, basil, eggs, mozzarella, salt and pepper and mix well.

6. Preheat the oven to 400°.

7. Lay each piece of eggplant flat.

8. Place 2 tablespoons of the ricotta filling on each piece and roll the eggplant.

9. In a casserole dish, put 1 cup of marinara sauce to evenly coat the bottom of the dish.

10. Arrange the eggplant rolls uniformly in the pan.

11. Top each roll with a little marinara sauce and place the dish in the oven.

12. Allow the eggplant rolls to cook for about 10 minutes or until warm and oozing in the center.

"It may take a great deal of courage, but unless we are willing to commit our actions to our beliefs, we may never realize how great an impact our efforts might make in improving the quality of life for others."
Samantha McGarity

Hungarian Chicken Paprikash with Gnocchi (Nokedli)

The City of Charleston Fire Department is an ISO Class 3 (Class 3/9) fire department consisting of 19 fire companies located throughout the city of Charleston, South Carolina. The Charleston Fire Department covers several areas including - Downtown "The Peninsula" West Ashley, James Island, Daniel Island, and John's Island.

Paprikash:
1 whole chicken or
8 thighs (the thighs are more cost effective and easier to portion, and you can avoid the argument of who will get the breast)
1 large onion
1 head of garlic
1-2 green peppers
1 large tomato
at least 2 Tbsp. of high quality red paprika (preferably Hungarian)
salt and pepper to taste

Gnocchi:
16 oz. of flour (or more, see the recipe)
1 large egg
16 oz. of sour cream
salt

1. Chop the onions (slice, dice, quarter, whatever way you like). Slice both the tomato and peppers, and peel the garlic cloves. Heat a few tablespoons of your favorite "grease" (oil, lard, butter) in a large enough pot to fit all the above, and dump the vegetables. Go crazy and drop a few peppercorns in there too. Add the paprika before it gets too hot to avoid burning it, otherwise it becomes bitter. Sauté it for a few minutes and then add the chicken. If you use a whole chicken you need to cut it up into peaces. Add enough water (or any stock you like) just to cover the dish, salt and pepper and bring it to boil. After it bubbles once, reduce the heat to simmer, and cook it until the meat is tender (approximately 45 minutes to an hour depends on the amount of the meat). Do not cover it! When it's done, cover the pot and let it stand for 10-15 minutes, so the meat can absorb some of the juices.

The Gnocchi:
1. Dump the sour cream in a bowl, crack the egg, salt to taste and start to add flour. Keep mixing it until it stops sticking to your hand. It should not be too soft or too hard.
2. Fill a large pot with water, add salt and bring it to boil. There are several ways to "rip" the dough into the water: get a small cutting board, wet it, rest it on the side of the pot, put a handful of the dough on it and start to cut approx. 1 inch pieces into the water with a knife. It takes some time to get efficient and fast to do it this way, and you'll figure out what size is the best for you. (I like large dumplings.) Get a handful of dough and pinch the same size as above, and drop it into the water. You won't feel your hand and fingers in just about 5 minutes.

Tips:
I do not measure the paprika but rather keep adding it until it is a beautiful dark red. I use the entire pepper, seeds, core and all. If you like it spicy, add one or two dried pueblos, jalapeños, or any of your favorite hot peppers. A gypsy friend told me, that he takes a small can of pickled cherry peppers (hot as you like), drain them, and toss them into the pot 10 minutes before he shuts off the heat. Delicious! You can add 8 oz of sour cream with 1 tblsp of flour mixed in it 10 minutes before it's done to add an extra body to it during colder days.

Charleston Fire Department
Fire Chef Zsolt Szoke, Charleston, South Carolina

Founded in 1670 as Charles Towne in honor of King Charles II of England, Charleston adopted its present name in 1783.

MEETING ST
BROAD

SQUEEZE
HAPPY HOUR
5:00-8:00PM
DAILY
$7 SPECIALTY MARTINIS
$6 SPECIALTY COCKTAILS
$6 WINE AND BUBBLES
$1 OFF ALL BEERS
CHARLESTON
FIRE
CHARLESTON
FIRE

CHARLESTON FIRE DEPARTMENT
CHARLESTON

Spaghetti Sauce

The Columbia Fire Department serves Columbia, the capital of South Carolina, as well as a 660 square-mile area of Richland County. The County's estimated 2009 population is 372,023, while the city of Columbia is estimated to have 129,333 residents. However, with the numerous and varied governmental agencies and local businesses within the Capital City, its daytime work force easily exceeds 400,000 persons.

olive oil or cooking oil
1 medium onion, chopped fine
6 or 7 garlic pods, chopped fine
1 stalk celery (optional)
½ to 1 bell pepper, chopped
oregano
sweet basil
2 or 3 bay leaves
crushed red pepper (optional, but use just a little)
grated cheese (Parmesan, Romano, or mix)
1 tsp. sugar
salt
pepper
40 oz. tomato sauce
1 small can tomato paste
1 or 2 tomatoes, chopped
water

1. Sauté onion, garlic, celery and bell pepper in oil until onion turns transparent.

2. Add tomato paste and brown slightly, stirring continuously.

3. Add tomato sauce, water.

4. Add all other ingredients and simmer slowly at least two hours.

For Meat Sauce:
1. Brown 1 pound ground beef.

2. Drain. Salt and pepper meat until it is slightly over-seasoned and add to sauce.

3. Omit salt and pepper from sauce, the seasoning in the meat will be enough.

South Carolina Fire Academy
Firefighter AJ Esposito, Columbia, South Carolina

"That's the life, being a fireman. It sure beats being a ballplayer. I'd rather be a fireman."
Ted Williams - Boston Red Sox 1940

GRANNY MORRISON'S COBBLER

The Greenville City Fire Department was established in 1876 with the mission of saving lives and protecting property. The Fire Department of today performs a variety of duties. We still fight fire, but we also respond to other emergency situations including vehicle extrications, medical emergencies, hazardous materials incidents, high-angle rescue, structural collapse, swift water rescue, and any natural or man-made disasters.

3 C. sweetened to taste fruit with natural juice (fresh or frozen - if frozen, thaw to room temperature)
blackberries
sliced peaches
sliced apples
pitted cherries (if using apples or cherries, sprinkle with ¼-½ C. sugar and heat until juice appears blackberries or peaches just sprinkle with sugar and stir juice will naturally appear fairly quickly without heating)
1 stick margarine, melted
¾ C. self-rising flour
¾ C. sugar
½ C. milk

1. Preheat oven to 375 degrees.

2. Place fruit with juice in 1½ quart ovenproof dish.

3. Add melted margarine to fruit. Mix flour, sugar and milk until smooth.

4. Pour batter over berry mixture, do not stir it in. Bake at 375 for about 45 minutes or until brown.

5. About halfway through baking (when batter has started to set) dot top with margarine (a couple of pats) cut into small pieces and then sprinkle the top with 2 tablespoons sugar. Let cool at least 15 minutes to let cooked batter soak up excess juice.

6. Serve warm or at room temp. Ice cream a plus.

GREENVILLE CITY FIRE DEPARTMENT
Fire Specialist James Cantrell, Greenville, South Carolina

God watches out for fools and firefighters, because somedays He can't tell us apart.
Jeff Smith, Galveston Fire Department

Bacon-Wrapped Chicken with Chile Cheese Sauce

The Brookings Fire Department is comprised of 48 members - three career members and 45 dedicated and professional volunteers. We protect the City of Brookings, 186 square miles in Brookings County and 27 square miles in Moody County.

8 slices bacon
4 boneless, skinless chicken breast halves, cut in half lengthwise (making 8 long pieces)
1 (10.75-oz.) can condensed cheddar cheese soup
⅓ C. milk
1 tsp. chopped canned chipotle peppers or bottled chipotle hot pepper sauce (or to taste)
dash Worcestershire sauce

1. Preheat oven to 400 degrees F. Line a rimmed baking sheet with aluminum foil; set aside.

2. Wrap a slice of bacon around each piece of chicken and place on the baking sheet. Bake about 20 minutes or until cooked through. Remove chicken from the oven. Preheat the broiler.

3. Meanwhile, in a saucepan, bring the soup, milk, chipotle pepper and Worcestershire sauce to a simmer over a medium heat. Turn off heat, cover to keep sauce warm.

4. Broil the chicken 4 to 5 inches from heat for 1 to 2 minutes or until the bacon is really sizzling. Serve chicken topped with cheese sauce.

Brookings Fire Department
Fire Chief Darrell Hartmann, Brookings, South Dakota

The only equipment available to fight the London fire in 1666 were two-quart hand syringes and a similar, slightly larger syringe; it burned for four days.

Fajita Grilled Steak

Both county and the city were named after one of South Dakota's pioneer promoters, Wilmot Wood Brookings (1830 - 1905). Brookings set out for the Dakota Territory in June 1857. He arrived at Sioux Falls on August 27, 1857, and became one of the first settlers there. After a time in Sioux Falls, Brookings and a companion set out for the Yankton area. The two soon encountered a blizzard that froze Brookings' feet which both had to be amputated.

1 small white onion, chopped coarsely
1 garlic clove, peeled and roughly chopped
1 Tbsp. fresh lime juice
½ tsp. cumin
¼ tsp. cayenne
½ tsp. salt
1 lb.beef skirt steak or flank steak, trimmed of surface fat

1. In a food processor or blender, combine one-quarter of the onion, the garlic, lime juice, cumin and salt. Process to a smooth paste.

2. Place the skirt steak in a non-aluminum baking dish.

3. Using a spoon, smear the marinade over both sides of the skirt steak.

4. Cover and refrigerate for at least 1 hour or up to 8 hours. (not more than 8 hours)

5. Remove the steak from the marinade.

6. Oil the steak well on both sides and lay it over the hottest part of the grill.

7. Grill, turning once, until richly browned and done to your liking, about 1½ to 2 minutes per side for medium-rare.

Brookings Fire Department
Fire Chief Darrell Hartmann, Brookings, South Dakota

The London fire stimulated the development of a two-person operated piston pump on wheels.

Lasagna Wrapped Smoked Sausage

Brookings is a city in Brookings County, South Dakota, United States. Brookings is the fourth largest city in South Dakota, with a population of 22,056 at the 2010 census. It is the county seat of Brookings County, and home to South Dakota State University, the largest institution of higher education in the state.

smoked sausage
lasagna noodle
1 Tbsp. oil
spaghetti sauce
mushrooms
peppers
onion diced up
mozzarella cheese

1. Preheat oven to 350 degrees.

2. Slice a smoked sausage up in pieces the width of a lasagna noodle. This will help determine how many lasagna noodles you will need to cook up. Bring water to a boil and place noodles in the pot with a tablespoon of oil to keep them from sticking together. Add a few extra noodles in case one of them tears apart.

3. Prepare your favorite jarred spaghetti sauce in a pan with any extras you might desire. (For example, I add mushrooms, peppers, and onion diced up. Cook this over a low flame while the noodles are cooking.)

4. Slice each piece of sausage down the middle and place a slice of mozzarella cheese in the center of each smoked sausage piece. When the noodles are cooked, drain and proceed to wrap each sausage piece into a noodle. You may place a long toothpick in them to seal them closed.

5. Place the wrapped sausages in a 9 x 12 x 13 pan and pour the spaghetti sauce over top. Place slices of the mozzarella cheese over top and bake at around 350 degrees for one hour.

Brookings Fire Department
Fire Chief Darrell Hartmann, Brookings, South Dakota

Tough times don't last; tough people do! Stay low, move fast.
Jon Cummings, FDNY, Manhattan

Chinese Coleslaw

In 1874 Major General George Armstrong Custer led an expedition into the Black Hills during which gold was discovered in French Creek, 13 miles south of Hill City. The discovery of gold opened the Black Hills, and the Hill City area, to mining. Hill City was first settled by miners in 1876 who referred to the area as Hillyo. This was the second American settlement in the Black Hills. Hill City is the oldest city still in existence in Pennington County.

1 bag of coleslaw
1 pkg. of beef flavored Ramon noodles - broken into small pieces (the flavor packet is used in the dressing)
½ C. of toasted slivered almonds

Dressing:
½ C. oil
½ C. sugar
½ C. vinegar (I prefer apple cider vinegar)
flavor packet from Ramen noodle packet

1. Mix together in large bowl, 1 bag of coleslaw, 1 package of beef flavored Ramon noodles - broken into small pieces (the flavor packet is used in the dressing) and ½ cup of toasted slivered almonds.

Dressing:
1. Shake dressing ingredients together and pour over ingredients in bowl and serve.

I shall never lead you into a situation we cannot overcome. Our job is to train and be prepared for any situation.
Harry Wood

BUTTON'S SWEET & SOUR PORK

Sioux Falls is a great community to live and raise a family. We consider our fire stations part of the neighborhood. Sioux Falls Fire Rescue is one of less than 200 fire departments in the country that is nationally accredited. We have 194 members and 10 fire stations strategically placed throughout the city.

4 lbs. boneless pork loin, cubed
3 green peppers, chunked
5 C. cooked Uncle Ben's Rice
2 cans pineapple rings
6 eggs
2 C. flour
salt
pepper
corn starch

Sauce:
4 C. sugar
8 Tbsp. soy sauce
3 C. ketchup
1½ C. pineapple juice (In addition to the juice from the pineapple rings)
4 C. white vinegar

1. In a bowl, whip eggs. Heat ¼ inch oil in a skillet. (Use a wooden match to determine proper oil temperature.)

2. Coat pork in eggs and flour; fry in oil. In a saucepan, mix sauce ingredients and bring to a boil.

3. Reduce heat, add a mixture of cornstarch and water to thicken sauce.

4. Add meat, pineapple pieces and green peppers to sauce.

5. Serve over rice.

SIOUX FALLS FIRE RESCUE
Captain Jon Button, Sioux Falls, South Dakota

When a group of insurance companies in New York had a self-propelled engine built in 1841, the firefighters so hindered its use that the insurance companies gave up the project.

CHICKEN ENCHILADAS

As one of the largest fire departments in the upper Midwest, we are committed to the safety of Sioux Falls. We do more than fight fires, as emergency medical calls now make up more than half our emergency responses. We also have one of the larger Public Access Defibrillator programs in the country, and regularly provided CPR classes.

4 lbs. chicken breast
2-3 packets taco seasoning
1 family size, one regular cream of chicken soup
24 oz. sour cream
2-4 cans diced green chiles
cholula hot sauce
8-12 Tortillas
1 pkg. Mexican or Colby Jack shredded cheese
1 bag tortilla chips

1. Mix the cream of chicken, sour cream, and diced green chilies together, adding about 2 tablespoons of Cholula.
2. Dice the chicken after removing as much fat as possible, and fry it with a little olive oil.
3. Once the chicken is cooked thoroughly, add the taco seasoning with the appropriate amount of water.
4. Once the seasoning has settled and mixed, add half of the cream of chicken mix, and stir until hot and mixed.
5. Add 2 spoonfuls of chicken enchilada stuff to tortilla, roll, and place in baking pan. Continue until all tortillas are rolled and ready.
6. Pre heat oven to 350-375.
7. Pour the remaining mix of cream of chicken on top of the rolled enchiladas, spread evenly, and sprinkle the shredded cheese over the sauce, until it's completely covered in cheese.
8. Place in oven, for 30-45 minutes; once cheese is melted and bubbling, they are ready.
9. Any remaining chicken enchilada mix that doesn't have room to go into the tortillas makes an excellent chip-dip to snack on while you wait for the oven portion.
10. A side of rice or refried beans compliments the enchiladas, and the chips help as well.
11. For spicier enchiladas, add diced jalapenos and/or increase the amount of Cholula; Tapatio hot sauce works also. Enjoy!

SIOUX FALLS FIRE RESCUE
Captain Jarud Neises, Sioux Falls, South Dakota

Low-fat Crab Rangoon

The community is a much needed partner in helping one another, so we are tasked with providing public EMS education and other community wide safety programs. The V.L. Crusinberry fire training center located on the Airport grounds is a regional center for Firefighter training and provides a solid training environment and curriculum.

8 oz. pkg. imitation crab
12 oz. weight watchers whipped cream cheese with onion and chives
½ C. fat free sour cream
2 tsp. soy sauce
2 tsp. Worcestershire Sauce
1 pkg. wonton wrappers

1. Preheat oven to 350.

2. Finely chop imitation crab meat.

3. Mix crab with soy and Worcestershire sauces, cream cheese and sour cream. Mix until you have a consistent creamy mixture.

4. Put teaspoon size ball of cream into the middle of wonton wrappers.

5. Wet the edges of the wontons to get them to stick together. Fold the wontons up together pinching wetted edges together.

6. Place on greased baking sheet. Spray all the Rangoon's with cooking spray.

7. Bake for 12 – 15 minutes.

8. Edges of wonton may look burned, this is normal.

9. Serve with your favorite Stir-fry dish.

Sioux Falls Fire Rescue
Firefighter Jason McManigal, Sioux Falls, South Dakota

The aerial ladder wagon appeared in 1870; the hose elevator, about 1871.

Powers' Zucchini Cake

The first documented visit by an American was by Philander Prescott, who camped overnight at the falls in December 1832. Captain James Allen led a military expedition out of Fort Des Moines in 1844. Sioux Falls, South Dakota, began as a city in the mid-1850s. As the town grew, a fire company was needed and in 1885 the first hook and ladder company was created

½ C. butter
½ C. oil
1¾ C. sugar
2 eggs
1 tsp. vanilla
½ C. sour milk (add 1 tsp. vinegar to milk to sour)
½ tsp. baking soda
2½ C. flour
4 Tbsp. cocoa
2-3 C. grated zucchini
chocolate chips
chopped nuts, optional

1. Preheat oven to 350.

2. Cream together the butter, oil, and sugar.

3. Add eggs, vanilla, sour milk, baking soda, flour, cocoa and zucchini.

4. Spread in a greased and floured 9 x 13 pan.

5. Sprinkle with sugar, chocolate chips and nuts (optional).

6. Bake at 350 for 40-45 minutes.

Captain Jim Powers, Sioux Falls, South Dakota

Advice to crew: "If everything seems to be going well, you obviously have overlooked something."
Cpt. Edward Moss, Bowling Green, Ky.

Salmon Patties

The Memphis Fire Department's Emergency Medical Service system is one of the first-based EMS systems in the United States beginning in 1966 with a fleet of eight emergency ambulances that handled approximately 8,000 runs during its first year of service.

2 (16 oz.) cans salmon, drained and boned
6-eggs
medium size onion, diced
½ C. flour, self rising
8 oz. saltine crackers, crumpled

1. Mix all ingredients.

2. Can use some of the drained salmon liquid to keep patty solidified.

3. Medium heated pan or skillet shortening or preferred oil.

4. Make a hamburger size patty. Cook till golden brown on both sides 3 or 4 min per side.

5. Can add more flour, or crackers to taste.

To a truckie there are only two types of glass: glass that is broken and glass that is about to be broken.
Dave Houseal, Harrisburg FD

FIREHOUSE BEANS

Freeport has a rich history. In 1528, Cabeza de Vaca landed in the area and named the river "Los Brazos de Dios." In 1822, Stephen F. Austin landed at the mouth of the Brazos River and founded Velasco. In the next 15 years, about 25,000 people entered the Republic of Texas through Velasco. In 1836 following the decisive battle of San Jacinto, Velasco was made the first capital of the Republic of Texas by interim President David G. Burnet. In 1929, the river was diverted south of town, leaving the Old Brazos riverbed as a protected harbor leading to the Gulf of Mexico. Originally two towns, Velasco & Freeport, on opposites of the Old Brazos River, joined to become the City of Freeport in 1957. Freeport is a part of the Texas Independence Trail.

2 large jars of Great Northern White Beans
1½ C. of sugar
1½ C. of brown sugar
2 sticks of butter, melted
1 lb. of bacon, cooked and crumbled
1 small to medium onion, chopped

1. Mix all of the ingredients in a 9 x 13 glass baking dish, bake uncovered at 300 degrees for two to three hours (based on preference).

Option B:
1. 3 C of sugar and no brown sugar.

Option C:
1. Drain one or both of the jars of beans for a dryer finished dish.

A fireboat, not limited to hydrant supply, can deliver as much as 10,000 gal per min.

Barbecue Chicken Legs

The Round Rock Fire Department is made up of professional men and women dedicated to your safety and quality of life. We are members of your community, neighbors and friends who are here to ensure that you and your property are protected from harm. We are able to do this by providing exceptional service and striving to meet internal and external goals. We are 129 individuals who work as a team to operate seven fire stations 24 hours a day, 365 days a year.

18 whole chicken legs
1 Tbsp. canola oil
¼ whole onion, diced
2 cloves garlic, minced
1 C. ketchup
¼ C. packed brown sugar
2 Tbsp. (additional) brown sugar
4 Tbsp. distilled vinegar (less to taste)
1 Tbsp. worcestershire sauce
⅓ C. molasses
4 Tbsp. chipotle adobo sauce (the adobo sauce chipotle peppers are packed in)
dash of salt

1. Preheat oven to 425 degrees. Place chicken legs on a broiler pan or any pan with a rack and then roast for 20 minutes.

2. While the chicken is roasting, heat canola oil in a saucepan over medium-low heat. Add onion and garlic, and cook for five minutes, stirring, being careful not to burn them.

3. Reduce heat to low. Add all remaining ingredients and stir. Allow to simmer while the chicken roasts. Taste after simmering and add whatever ingredient it needs (more spice, more sugar, etc.)

4. After 20 minutes of roasting, crank on the broiler to get a little color on the legs. Broil for five minutes.

5. Remove chicken legs from oven. Reduce oven to 350 degrees. With tongs, dip each chicken leg into the sauce, submerging completely. Place back onto the pan. After all chicken is coated, return pan to oven for five minutes, or until hot and sizzling.

6. Remove from oven. Brush/dab generously with remaining sauce. Allow chicken to sit a few minutes before serving.

ROUND ROCK FIRE DEPARTMENT
Captain Allen Duennenberg, Round Rock, Texas

Even before the US Civil War, barbecue was a popular meal choice. It was originally used to describe the smoking of whole pigs or steers over open fires. People would gather from miles away just to taste the delicious barbecued meat.

Cowboy Pot Luck

The Round Rock Fire Department is making great strides in becoming the Central Texas hub for Firefighter training. Not only is our staff of dedicated Firefighters continually going through training at our facility, we offer Texas Engineering Extension Service (TEEX) classes, as well as other training classes that are designed for continued education of the highest caliber.

3 lbs. of beef hamburger
1 lb. of bacon (chopped)
2 large cans of bake beans
1 C. of barbecue sauce
12 oz. of water or
12 oz. of beer
shredded cheese
uncooked biscuit

1. Take 3 pounds of beef hamburger, one pound of bacon (chopped) and mix with 2 large cans of bake beans.

2. Add half of cup of barbecue sauce to mix with meat and beans. Place mixture in large baking dish or a throw away roasting pan.

3. Preheat oven to 375 degrees and place mixture inside for one hour until hamburger, bacon and beans completely together.

4. Drain excess render and fat to reduce. After reduction, you may add 12 ounces of water or 12 ounces of beer to assist in moisture.

5. When mixtures are completely cooked, then add shredded cheese of choice over entire mixture and melt cheese.

6. Add uncooked biscuit on top of cheese and let biscuits bake until done. You may have to increase heat to bake biscuits completely.

7. When biscuit have baked; pull dish out and serve. (I cook this with creamy garlic mash potato.)

8. You first add potatoes to plate or bowl; then add the Cowboy Pot Luck on top. If cheese is what you like then add sour cream to make mixture creamier. Now you have completed your meat. Potato, cheese and bread dish.

Firefighter Bruce Allamon, Round Rock, Texas

"The possible, firefighters do fast -- the impossible just takes time".
unknown

Cream Cheese Cinnamon Rolls

The Round Rock Fire Department Hazardous Materials Team is comprised of 21 firefighters and one Team Leader. Seven of these team members are on each shift, with the majority of the team housed at Fire Station 6. The team has a trailer that carries all of their supplies in the event of an incident.

¼ C. warm water
¼ C. butter, melted
½ (3.4 oz.) pkg instant vanilla pudding mix
1 C. warm milk
1 egg, room temperature
1 Tbsp. white sugar
½ tsp. salt
4 C. bread flour
1 (.25 oz.) package active dry yeast or 2 tsp. of bread machine yeast or any bread yeast

Filling:
¼ C. butter, softened
1 C. brown sugar
4 tsp. ground cinnamon
¾ C. chopped pecans

Icing:
½ (8 oz.) pkg. cream cheese, softened or creamier icing use 1 (8 oz.) pkg. of cream cheese
¼ C. butter, softened
1 C. confectioners' sugar
½ tsp. vanilla extract
1½ tsp. milk

1. In the pan of your bread machine, combine water, melted butter, vanilla pudding, warm milk, egg, 1 tablespoon sugar, salt, bread flour and yeast. Set machine to Dough cycle; press Start.

2. When Dough cycle has finished, turn dough out onto a lightly floured surface and roll into a 17 x 10-inch rectangle. Spread with softened butter. In a small bowl, stir together brown sugar, cinnamon and pecans. Sprinkle brown sugar mixture over dough.

3. Roll up dough, beginning with long side. Slice into 16 one-inch slices and place in 9x13-buttered pan. Let rise in a warm place until doubled, about 45 minutes. Meanwhile, preheat oven to 350 degrees F (175 degrees C).

4. Bake in preheated oven for 15 to 20 minutes. While rolls bake, stir together cream cheese, softened butter, confectioners' sugar, vanilla and milk. Remove rolls from oven and top with frosting.

Round Rock and Williamson County have been the site of human habitation since at least 9,200 BC.

Deer Camp Chili

Members of the Round Rock Hazardous Materials Team participate in a total of 36 drills in conjunction with the Williamson County Hazardous Materials Team and the 6th civil Defense Team. 33 percent of the Hazmat Drills that take place in Williamson County are hosted in Round Rock. These drills allow the team to exercise their skills and allow them to be prepared in the event of a situation that requires their expertise.

3 lbs. of ground beef hamburger meat
chopped white and yellow onions
1 Tbsp. of salt
black coarse pepper
1 Tbsp. of crush red pepper (optional)
2 Tbsp. of garlic powder
½ C. of fresh cilantro (chopped)
½ C. of celery (chopped).
¼ C. of red cider
¼ C. of Worcestershire sauce
2 Tbsp. of cumin
2 Tbsp. of red chili powder for seasoning
2 Tbsp. of dry mustard.
cornstarch or couple of spoons of flour
6 small cans of tomato sauce
2 cans of beer

1. Take 3 lbs. of ground beef hamburger meat (preferably) 70/30 meat ratio.

2. If you are cooking with deer hamburger (50/50) is ideal to complete the recipe. Cook meat on medium heat with chopped white and yellow onions.

3. Cook until meat has render and drain the fat.

4. Then cook meat and onions on low heat and add tablespoon of salt, black coarse pepper, table spoon of crush red pepper (optional), two table spoons of garlic powder, half of cup of fresh cilantro (chopped), and half a cup of celery (chopped).

5. Cook until meat and vegetables are binding together. Add quarter cup of red cider vinegar and quarter cup of Worcestershire sauce and mix chili together for 20 minutes.

6. Add two tablespoons of cumin, two table spoons of red chili powder for seasoning and two tablespoons of dry mustard.

7. Cook for additional 15 minutes to bind all spices together. Add 2 cans of Rotel sauce to mixture.

8. Add 6 smalls can of tomato sauce with two cans of beers. Cook slowly until chili is thoroughly cooked.

9. If chili is thin, add cornstarch or couple of spoons of flour to thicken mixture. If additional seasoning is needed; simple add for taste.

10. If you have notice I do not cook chili with beans. If you like beans in your chili; then add you bean of choice.

Firefighter Bruce Allamon, Round Rock, Texas

"We bravely march into the depths of hell to face the fears of many, and extinguish the demons that rage free".
Anonymous

Truck 1's Rib BBQ Dinner

The Round Rock Fire Department's Rock Solid Team not only performs at Round Rock's 18 elementary schools every year, it frequently works with fire departments in other cities to help them develop similar programs and has conducted seminars for the national convention of the International Association of Fire Chiefs. Since 1994, the Team has reached an estimated 67,000 elementary children.

4 racks of baby backs
2 bottles of Wishbone Red Wine Vinaigrette
Fiesta Brand Chicken Fajita seasoning
2 foil turkey roasters with lids

3 C. of cooking rice (You can use Uncle Bens if you don't want the good real rice) the rice to water ratios will change with Uncle Bens or HEB rice.
6 C. of water

2-16 oz cans of Bushes baked beans

blackberry pie and vanilla ice cream
2 Pillsberry pie crusts
2 (21 oz) cans of Lucky leaf blackberry pie filling
½ gallon of Blue Bell vanilla ice cream

1. Rub down ribs with fiesta brand chicken fajita seasoning (Make sure it's the one with salt, lemon, and butter flavorings). Place the coated ribs in the 2 roasters and in the fridge.

2. Get BBQ pit going and keep temperature around 350 degrees. We like to use Mesquite wood.

3. While the BBQ pit's getting ready, start making pie. Roll out 1 pie dough on a lightly floured surface. Place dough in a pie plate. Open both cans of Lucky leaf blackberry pie filling and pour into pie plate. Lattice the other pie crust on top. Crimp the edges. Paint a light coating of milk on lattice top and sprinkle with sugar. Bake at 350 for 45-50 minutes. Keep an eye on your oven-they do not cook all the same. If the edges start to brown too much, cover them with a little foil.

4. Take the ribs out of the roasters and place them onto the BBQ pit. Just brown/smoke those good on both sides, take them off, and place them back into the roasters and cover them with the red wine vinaigrette. 1 bottle for each roaster. Bake at 350 in the oven for 1 hour. The meat should fall off the bone.

5. On the stovetop-start the 6 cups of water to boiling. Put the 3 cups of rice in and simmer with lid on for 15 minutes. Do not take the lid off to look at it.

6. In another pot-start the beans heating. You can just barely turn the stove on to simmer.

Captain Allen Duennenberg, Round Rock, Texas

This recipe will work for chickens or steaks too. This is a sure fireman pleaser. The preparation time is figured and everything should fall into order.

JOHN-BA-LAYA

Looking back over the past five years, the North Davis Fire District has done all it set out to do, and more, by providing excellent fire and emergency response services for both Clearfield and West Point. The District now has two fire stations, one in Clearfield and its headquarters in West Point, as well as 18 full-time firefighters/EMT's, plus Fire Chief Roger Bodily, Fire Marshall Craig Whiteside, and District Clerk Michelle Marsh, and 14 part-time firefighters/EMTs.

vegetable oil (enough to cover the bottom of a large cast iron enameled dutch oven
hillshire farms summer sausage (the cajuns use andouille but we can't find any in utah) slice it crosswise about □ inch thick
3 medium onions chopped
2 bell peppers (if available I will use one green and one red or yellow) the red or yellow give it some color
2 or 3 cloves of garlic (depends on how much garlic you like)
3 boneless/ skinless chicken breasts cut into chunks
1 ham steak cut up into bite size chunks
3 C. of medium grain white rice
6 C. of water
at least a pound if not more of green shelled and veined shrimp
1 bunch of green onions chopped
about half of a bunch of parsley chopped
salt and pepper
cayenne pepper
3 or 4 bay leaves
oregano to taste
dried thyme

1. Get the oil in your Dutch oven hot and then add the sausage, onions, and bell peppers, about a teaspoon of salt and about a teaspoon of cayenne pepper. Let these items cook over medium heat until the veggies get brown and soft.

2. Stir often so that they don't burn. Maybe about 10 to 15 minutes.

3. Season the chicken with salt, pepper and more cayenne pepper then add them to the pot with the garlic, ham, bay leaves, thyme and oregano.

4. Cook and stir often until the chicken is well browned. About 10 minutes.

5. Stir in the rice and get it evenly distributed into the mixture then slowly add the water.

6. Cover and cook on medium leave for about 30 to 35 minutes, but do not stir or open the lid during cooking.

7. Add the shrimp, recover and let stand for about 5 minutes until the shrimp is done.

8. In a large serving bowl put the green onions and parsley in the bottom and then pour the jambalaya out of the cast iron pot into the mixing bowl.

9. Stir the parsley and green onions into the rice mixture until evenly distributed. Keep and eye out for the bay leaves and remove them as you find them.

A number of years ago I took my family to Disneyland and we had dinner one night at the restaurant next to The Pirates of The Caribbean ride. On the menu was Jambalaya. I had never tried Cajun food but was intrigued to try something adventurous. I ordered the jambalaya and fell in love with it. After returning home to Utah I tried to find recipes for Jambalaya that I liked.

NORTH DAVIS FIRE DISTRICT
Captain John C. Taylor, Clearfield, Utah

Betty's Ham Loaf

For the first 36 years of its existence, Salt Lake City relied on the services of volunteer firefighters to protect lives and property. The volunteers strove to uphold their motto, "We Aim to Aid and Work to Save." By the 1880s this system was insufficient for the growing city, and a professional force was created. A terrible fire in the early morning hours of June 21, 1883, proved to be the final factor in the establishment of a professional fire department.

1 lb. cured ham chopped finely
1 lb. ground pork
1 C. milk
2 C. bread crumbs
2 eggs
½ tsp. white pepper

For Glaze:
1 C. brown sugar
½ C. vinegar
½ C. water
1 tsp. dry mustard

1. First, preheat oven to 350. Mix all ingredients for the glaze by whisking together and use this to baste the ham loaf when it is in the oven.

2. Mix bread crumbs with the pepper, then add milk and eggs and mix completely. In a bowl put the chopped cured ham and ground pork, mixing them thoroughly, then add the bread crumb mixture to the meat and mix them completely. Place on a pan in either one large loaf or two smaller loaves, with a groove in the middle about ½ inch deep to catch the glaze, and bake 1½ hours.

3. Baste the ham loaf every ½ hour, and the glaze will form a nice crisp coating as well as give flavor with the mustard in it. When finished, remove from oven and serve in ½" slices.

4. With the oven being on you can make macaroni & cheese to go with it or baked potatoes, and apple sauce always goes great with pork.

SALT LAKE CITY FIRE DEPARTMENT
Captain Paul Runyon, Salt Lake City, Utah

The first modern standards for the operation of a fire department were not established until 1830, in Edinburgh, Scotland.

Battalion Brussels Sprouts

Fire Station Three is one of the oldest operating firehouses in Vermont. The original Fire Station near this site burned down, leading to the decision to establish a career Fire Department. In 1895, the Burlington Fire Department was founded and this station was placed in service.

8 strips of thick cut apple wood smoked bacon
2.5 lbs. fresh Brussel sprouts
1 small yellow onion
3 cloves fresh minced garlic
1 Tbsp. dried basil
1 tsp. black pepper
2 tsp. kosher salt

1. Preheat oven to 375 degrees F

2. Trim stems and remove outer leaves of each sprout.

3. Cut sprouts in half and place in a large mixing bowl.

4. Dice the onion into □ inch pieces. Cook the bacon in a fry pan until crispy (reserve the drippings). Take the bacon out of the pan and set aside.

5. Take the bacon drippings and pour them over sprouts (about 1/3 cup). Add the garlic, basil, black pepper, diced onion, and salt to the bowl.

6. Mix all together until the sprouts are coated. Pour sprouts onto a baking sheet keeping them to a single layer. Place baking sheet into the preheated oven for about 35 minuets or until sprouts are golden brown on the cut side.

7. Be sure to flip sprouts half way through cooking. Remove from the oven. Crumble the bacon over the sprouts and serve. Enjoy!

Burlington Fire Department
Senior Firefighter Chris Laramie, Burlington, Vermont

But aloud the praises, and give the victor-crown.
To our noble hearted Firemen, who fear not danger's frown.
Fredric G. W. Fenn

LESLIE PEARÇE'S CHOCOLATE ÉCLAIR CAKE

In 1862, the American Civil War came to Franklin, in what was referred to as the Joint Expedition Against Franklin. As several United States Navy Flag steamships, led by the USS Commodore Perry, tried to pass through Franklin on the Blackwater River, a band of local Confederates opened fire on the ships. As stated by an officer aboard one of the ships, "The fighting was the same—Here and there high banks with dense foliage, a narrow and very crooked stream, with frequent heavy firing of musketry."

1 box graham crackers
2 small boxes vanilla instant pudding
3 C. milk
1 (8 oz) ctr. cool whip
1 can chocolate frosting

1. Mix the milk and pudding together until no lumps are found, fold in soften cool whip.

2. Scoop enough of the pudding mixture to lightly cover the bottom of a 9 x 13 dish. Then, layer Graham crackers on top and repeat two more times ending with crackers on top.

3. Microwave the frosting for about 20 - 30 seconds, just enough to make it easy to pour on top of crackers. Spread it evenly covering the Graham crackers. Let it set for at least 2 hours, in the refrigerator.

The Franklin Fire & Rescue Department employs 15 career staff and approximately 50 active volunteers and junior Fire Cadets.

Clown Crock Chili

The Virginia Beach Fire Department began as an all volunteer department in 1906 when the Town Council noted the need to provide fire equipment to protect the rapidly growing resort area. During the next 20 years, the volunteer department met the many challenges that it faced, but noted that 24-hour fire protection was desperately needed. In 1928, the town hired a paid staff of firefighters that also performed the duties of police officers. This dual duty system was necessary as the town did not have the funding to hire single function employees.

1½ lbs. beef cut in one inch cubes
½ lbs. hot italian sausage cooked and cut into bite sized pieces
½ lb. sweet italian sausage cooked and cut into bite sized pieces
1 large can (40 oz) kidney beans drained
1 can (28 oz) diced tomatoes do not drain
1 can (10 oz) Rotel tomatoes do not drain
1 can (15 oz) of corn drained
1 large bell pepper diced
1 large onion diced
1 clove garlic minced
4 oz beef stock
□ C. chili powder
1 Tbsp. black pepper
½ Tbsp. salt
½ tsp. crushed red pepper
½ Tbsp. cumin
olive oil, corn starch, hot sauce add to desired taste

1. In a large pan, cook sausage until slightly pink. Remove and set aside.

2. Coat bottom of large pan with olive oil. Add chopped onion, peppers, and garlic and sauté until soft approx 5 minutes.

3. Add cubed beef and brown; add sausage. Stir to combine; stir all together to a crock pot.

4. Add drained kidney beans and tomatoes; stir together.

5. Add approximately 4 ounce of beef broth. Set remainder aside to use with corn starch to thicken.

6. Add spices.

7. Simmer approximately 8 hours on low.

8. Mix corn starch and remaining broth to thicken as desired.

9. Garnish with diced scallions and shredded Monterey Jack cheese. Enjoy!

Virginia Beach Fire Department
Fire Chef Keith Arnold, Virginia Beach, Virginia

"It comes down to tactics.... I don't want to do anything first. I want to do seven things all at once."
Tom Brennan

White Bean Chicken Chili

You respond to a working fire. It's 21:00 and near-freezing outside; the building is a three-story apartment with occupants trapped. The fire extends from the first floor into an open attic area. Firefighters began their work early, and about an hour in, fire is under control, and the occupants are safe. So how do you rehab your people? With this White Bean Chicken Chili. “It is one of my family's favorite winter meals. Really good with fresh homemade bread.” says Chief Boyd.

- 1 Tbsp. olive oil
- 1½ lb. boneless chicken breast
- ¼ C. chopped onion
- 1 can chicken broth
- 4 oz can chopped green chilis
- 19 oz can white kidney beans
- 2 green onions, chopped
- 1 tsp. garlic powder
- 1 tsp. ground cumin
- ½ tsp. oregano leaves
- ½ tsp. cilantro leaves
- □ tsp. ground red pepper
- onion and Monterey Jack cheese for garnish

1. Cut chicken breasts into small cubes.

2. Heat oil in large saucepan over med/high heat. Add chicken and onions, cook 4-5 minutes.

3. Stir in broth, chilies, spices, simmer 15 minutes.

4. Stir in beans, simmer 5 minutes

.

5. Top with onions, garnish with Monterey Jack cheese. Yields 4 cups.

Bellingham Fire Department

Former Fire Chief William Boyd, Bellingham, Washington

Back in the early days when cars first appeared on the roads, they were all black. So to help clear the roads when a fire truck was coming through, they painted them red so they would stand out.

Sausage Stuffed, Bacon Wrapped, Pork Tenderloin with Roasted Apples & Cider Sauce

Orcas Island Fire & Rescue members have been donating time to Patos Island under the guise of the informal Patos Island Fire Department since November 2005. The group has a membership of more than 60, and has donated well over 1000 hours towards nature habitat restoration, trail/campsite maintenance, building maintenance and brush abatement.

Filling:
2 Tbsp. unsalted butter
½ large yellow onion, chopped
3 garlic cloves, minced
1 lb. sausage, casings removed
1 Tbsp. fresh rosemary, chopped
1 medium Granny Smith apple, cubed
1 C. portobella mushrooms, roughly chopped
1 tsp. salt
¼ tsp. black pepper
4 oz. cream cheese, softened
½ C. sour cream
4 oz. feta cheese, crumbled

Pork
3½ – 4 lb. pork loin
16 slices bacon, I used 13, 6×7, depending on the size of your roast
4 medium apples, quartered, I used Fuji
3 Tbsp. unsalted butter
1 C. hard cider

Filling:
1. In a large skillet, over medium heat, melt butter. Add onion and cook until soft and lightly golden, about 5 minutes. Add garlic, sausage, rosemary and apples; cook, stirring occasionally. Stir in salt and pepper. Remove from heat and allow to cool completely.
2. In a small bowl, combine cream cheese, sour cream and feta; set aside.

Pork:
1. To butterfly, put pork loin on a work surface with short end facing you. Holding a long, thin sharp knife parallel to work surface and beginning along one long side, cut ½□ above underside of roast. Continue slicing inward, pulling back the meat with your free hand and unrolling the roast like a carpet, until the entire loin is flat. Cover with a sheet of plastic wrap. Using a meat mallet, lb.to an even thickness.
4. Uncover pork, spread cheese mixture evenly over loin, leaving a 1 inch border. Evenly distribute sausage mixture over cheese. Roll pork into a tight cylinder. Tie roast securely with kitchen twine in 1□ intervals. Wrap bacon lattice over roast.
5. Preheat oven to 400°. Place tenderloin in roasting pan. Scatter apples around roast. Add cider and water to pan. Roast pork until an instant-read thermometer inserted into center of loin registers 140°, about 1 hour to 1 hour and 15 minutes. Let roast rest for at least 20 minutes and up to 2 hours.
6. Place roast and apples on a serving platter.
7. Place the roasting pan with remaining bits and juices over medium heat. Add butter, hard cider, whisking to release any stuck bits from the pan. Stir occasionally for approximately 5 minutes. Season with salt and pepper. Strain into a serving bowl. Slice pork, and serve with apples and sauce spooned over top.

Patos Island Fire Department
Fire Chef Steven W. Siler, Patos Island, Washington

"I'm a secular priest ordained by training, experience, and, most importantly, the willingness to accept the mantle of command. That willingness encompasses the realization that failure is easy, and such failure could kill me or, worse kill someone else… "
Peter M. Leschak

CHICKEN/HAM/TUNA SALAD

Standing inconspicuously at 418 W. First Avenue in between Washington and Stevens in downtown Spokane, the Spokane Fire Department's original station No. 1 still stands over 122 years after its construction.

'meats such as chicken, ham, tuna, turkey, or even hot dogs

Salad Dressing:
miracle whip, mayo, or 'store brands'
mustard
sweet or dill pickle relish
bread & butter pickles diced or minced into a relish
celery, chopped/diced
green peppers and/or red peppers
onion

Other ingredients that could be added are:
green peppers, apples, mandarin oranges, grapes, hard boiled eggs, etc. that suit your tastes.

1. This 'salad' can be made with any 'meats', such as, chicken, ham, tuna, turkey, or even hot dogs that have been ground or chopped, although the 'smoked varieties' may not blend well with the sweetness of the other ingredients.

2. This basic dressing may be used as the main part of a number of 'salads' and can be adjusted to suit the main ingredient or the situation.

3. Quantities have been omitted and will depend on how many to be served. It is easier to make the dressing first and then add the 'dry' ingredients, mixing as you go. This is quick and easy to make for any number and the ingredients can be adjusted as preferred. It can be served over lettuce or as a sandwich with any type of bread or toast.

Fire-alarm systems came into existence with the invention of the telegraph. Today many communities are served either with the telegraph-alarm system or with telephone call boxes.

Jerry's Jumbo Chocolate Chip Cookies

Spokane's first residents were Native American. From the Spokanes, we get our name, which means "Children of the Sun." Spokane became an incorporated City on Nov. 29, 1881, encompassing 1.56 square miles. Back then, the City was known as Spokan Falls and had 350 residents. The "e" was added to Spokane in 1883, and "Falls" was dropped in 1891. The City suffered, perhaps, its biggest setback in 1889, when a fire ravaged downtown destroying 32 blocks.

5 C. of flour
2 tsp. baking soda
1½ tsp. salt

Mix In Separate Bowl (set aside):
2 C. (4 sticks) butter softened
1½ C. sugar
1½ C. brown sugar
4 tsp. vanilla
4 eggs
12 oz chocolate chips
12 oz M&M's

1. Whip butter until very smooth.

2. Add sugar and brown sugar.

3. Add eggs 1 at a time.

4. Add vanilla.

5. Mix until batter changes to a lighter color.

6. Slowly add flour mixture to batter.

7. Now add chocolate chips and M&M's by hand.

8. Place 6 cookies (approx. the size of a large ice cream scoop) on to a cookie sheet. Bake at 350 degrees for 15 to 22 minutes or until the edges of the cookies are turning brown.

Note: Do not over bake if you want softer cookies. If your cookies are turning out very flat you can add ¼ cup of flour, this will help.

Spokane Fire Department
Fire Chief Bobby Williams, Spokane, Washington

"If I yell at you while on scene don't take it personally, but if I yell at you off scene it's for a reason."
Your captain

Pasta Salad

The Operations Division is responsible for the delivery of emergency fire, rescue, EMS and HazMat response services to the City of Spokane. The Division is headed by the Deputy Chief of the Department who directly supervises 8 Battalion Chiefs, the Communications Manager and the Division Chief of Emergency Medical Services. The Department provides services with 17 front-line apparatus responding from 14 fire stations. Daily staffing is currently 58 personnel, including two on-duty Battalion Chiefs.

salad dressing [miracle whip, mayo, or 'store brands']
mustard
sweet or dill pickle relish
bread & butter pickles, diced or minced into a relish
celery, chopped/diced
green peppers and/or red peppers
onion
hard boiled eggs chopped
pasta

1. Cook the pasta as per the package instructions. Do not over cook! Pasta should not be completely cooked and should be rinsed with cold water to stop the cooking.

2. Allow to rest and refrigerate to chill. This allows better mixing and will prevent 'mushy' pasta.

3. Mix with pasta, garnish with sliced hard boiled eggs and paprika if desired.

4. Quantities have been omitted and will depend on how many to be served. It is easier to make the dressing first and then add the 'dry' ingredients, mixing as you go. This is quick and easy to make for any number and the ingredients can be adjusted as preferred. It can be served over lettuce or as a sandwich with any type of bread or toast.

I can think of no more stirring symbol of man's humanity to man than a fire engine.
Kurt Vonnegut

LADDER-4 STUFFED MEATLOAF

The Spokane Fire Department's Training Center is a 16,207 square foot building serving multiple roles. The building houses the Spokane Fire Department's training staff, Homeland Security and the SFD's "Channel 95" Broadcasting Studio and Audio/Visual Technical Services.

SPOKANE FIRE DEPARTMENT
Fire Chef Jerry Shaw, Spokane, Washington

2 eggs
⅓ C. milk
1 tsp. garlic
1 tsp. pepper
1 tsp. Worcestershire sauce
2 tsp. dried minced onions
2-3 slices of white bread crumbled
2-3 lbs. of lean hamburger
1 to 2 pkgs. of (Buddig) wafer sliced ham
1 small can sliced olives (optional)
1 small can sliced mushrooms (optional)
1 C. each of shredded cheddar and mozzarella cheese

1. Pre heat oven to 350 degrees.

2. Mix first 7 ingredients in a large bowl.

3. Add hamburger and mix with hands well.

4. Flatten hamburger on plastic wrap (about ½ to ¾ inch thick).

5. Place ½ of the (Buddig) wafer sliced ham on the hamburger.

6. Spread on the olives and mushrooms.

7. Sprinkle the cheese over the top of olives and mushrooms.

8. Place a second layer of (Buddig) wafer sliced ham on top.

9. Grab the edge of the plastic wrap and roll the hamburger into a loaf.

10. Seal up the ends and the edge of the hamburger to prevent insides from leaking out.

11. Bake 1½ to 2 hours, remove and let stand for 5-10 minutes.

12. Serve with baked potatoes and french bread.

For example, in the U.S. approximately 70 percent of all emergency medical calls are handled by the fire service.

TONY'S PULEHU BBQ SAUCE

On Sunday, August 4, 1889, fire destroys most of downtown Spokane Falls. It begins in an area of flimsy wooden structures and quickly engulfs the substantial stone and brick buildings of the business district. Property losses are huge, and one death is reported. Initially the fire is blamed on Rolla A. Jones, who was in charge of the water system and was said to have gone fishing after leaving the system in the charge of a complete incompetent. Later, city fathers will exonerate Jones, but this account, although false, will be repeated in many histories of the fire. Spokane will quickly rebuild as fine new buildings of a revitalized downtown rise from the ashes.

3 C. Tree Top apple cider
¼ C. light brown sugar
¼ C. molasses
¼ C. soy sauce
¼ C. ketchup
¼ C. apple cider vinegar
4 cloves garlic minced
2 Tbsp. fresh ginger grated
1 Tbsp. asian chili paste

1. Combine all ingredients in a sauce pot and bring to boil.

2. Lower temperature and reduce to half for marinade or to syrupy glaze or thicken with cornstarch for dip sauce.

3. Sauce at first boil is also great for braising.

FEO, Fire Equipment Operator. I drive the fire truck and operate all the equipment on it. My assigned truck is Pumper Ladder 13 (PL13) at Fire Station 13, Spokane Fire Department, City of Spokane, Washington.

ORCAS ISLAND
ORCAS

ORCAS
ORCAS
FIRE
ORCAS
KIMPLE

Healthy Spaghetti

Orcas Island Fire & Rescue was chartered in 1948. As San Juan County's second fire district, its role expanded from an all-volunteer service in the early 1970s to include emergency medical services, and a staff of Firefighter Paramedics, Firefighter EMTs, an EMT Administrative Assistant and the Chief , together with 60+ Volunteer Firefighters, EMTs and First Responders. OIFR also hosts the SJC Deputy Fire Marshal. The district established 24-hour staffing in 2005.

broccoli
ground beef, chicken, turkey, or pork
olive oil
garlic
tomato sauce
brown lentils
chili paste
1 Tbsp. tomato paste
balsamic vinegar
fresh parmesan
brown rice pasta

1. Start by adding far more veggies then meat, or even skip the meat all together. Before adding the veggies to the pan, strain the fluid but retain it. Caramelize your vegetables and then add the fluid back to the pan. Which vegetables to use really are up to you but you may want to stay away from very strongly flavored veg such as broccoli. It can be ground beef, chicken, turkey, or pork or a mixture of any of these. I brown my meat and vegetables in a small amount of olive oil.
2. Once everything is nicely browned you can add your herbs and spice. I usually save my fresh garlic for the very end of the browning process so it does not become burned. I wait until I add the tomato products because these already contain salt. Retain some of your fresh herbs to add ad the very end.
3. Here is my "secret" ingredient, brown lentils.
4. Now I start adding the tomato. You can use fresh if you like but it will be waterier and maybe not as flavorful out of season. To concentrate flavor and reduce moisture you can pre-roast them but for speed and ease, I just use a large can of organic diced tomatoes. Let the mixture simmer and reduce a bit.
5. Now you can add your tomato sauce. If you are looking to avoid all of those extra sugars and sodium, just add plain old tomato sauce.
6. After a little simmer time, this is the point you can begin to customize for your taste. Small additions can really bring a depth without adding a lot of calories. If you like spicy, add your favorite chili paste or concentrated sauce such as Sambal or Sriracha.
7. If you like your sauce to be extra "tomato-e" add a tablespoon or two of tomato paste or finely chopped reconstituted sun dried tomato. If your kinds just have to have that store bought sauce flavor, a tablespoon of catsup will do the trick. Just keep in mind it has loads of sugar. In this preparation, I added about a tablespoon of "Bragg" which is an amino acid preparation that tastes similar to soy sauce but with far less sodium and a nice dollop of balsamic vinegar.
8. Once the flavors are incorporated and the sauce has reduced to the consistency you like you are ready to go. I served my sauce over organic brown rice pasta. If you leave the cheese out of the sauce and sprinkle fresh parmesan over the top, a little will go a long way.

Division Chief/Firefighter-Paramedic Valerie Harris, Orcas Island, Washington

While doing research at UCLA Harbor Medical Center for a proposed new show about doctors, television producer Robert A. Cinader, happened to encounter "firemen who spoke like doctors and worked with them". This concept developed into the television series Emergency! about the profession called paramedicine.

Barbecued Baked Beans

The outlook for the fire service in West Virginia is extremely hopeful. We are looking to maintain a level of proficiency of our trade craft through training and the furthering of our educational goals and opportunities. We are steadfast proponents of the public's trust in the fire service and endeavor to maintain a balanced service, providing for the public while maintaining our fiduciary responsibility to the tax payers of West Virginia.

12-16 oz bacon
2 medium onions, thinly sliced
½ C. vinegar (white or apple cider)
1 C. ketchup
1 tsp. salt (or to taste)
1-2 tsp. dry mustard
½-1 tsp. curry powder (mild)
1 tsp. white pepper
1 tsp. celery seed
½ to ¾ C. packed brown sugar
hot sauce to taste (I use about 3 Tbsp. Texas Pete or Tabasco)
1 (16 oz) can kidney beans, drained
1 (largest size - 32 oz) can pork & beans, drained
1 (16 oz) can butter beans, drained

1. Cook all but 4 or 5 slices of bacon & drain; set aside cooked bacon. Saute onions in bacon grease until limp but not brown. Add vinegar, catsup, brown sugar, and spices. Cook over medium heat, stirring often.

2. Put drained beans in ungreased casserole dish. Crumble cooked bacon and mix in with the beans. Pour in the heated sauce; mix into the beans. Top with remaining slices uncooked bacon.

3. Bake at 350 degrees for one hour. Serves 8 to10 people.

NOTE:
Make this a day ahead of time if you can—it gives the flavors more of a chance to mix.

West Virginia State Fire Department
Firefighter Carol Nolte, West Virginia

West Virginia is one of two American states formed during the American Civil War (1861–1865), along with Nevada, and is the only state to form by seceding from a Confederate state

"Fire in the Hole" Hot Peppers & Sausages

Orange County, Virginia was formed in 1734. It included all areas west of the Blue Ridge Mountains, constituting all of present West Virginia. However, in 1736 the Iroquois Six Nations protested Virginia's colonization beyond the demarcated Blue Ridge, and a skirmish was fought in 1743. The Iroquois were on the point of threatening all-out war against the Virginia Colony, when Governor Gooch bought out their claim for 400 pounds at the Treaty of Lancaster (1744).

2 gallons hot peppers (mild banana peppers work best)
3-5 lbs. browned & drained Italian sausages, chopped
4 or 5 onions, sliced
2-36 oz bottles ketchup
2-12 oz bottles chili sauce
1 pint white or cider vinegar (I use cider vinegar)
3¾ C. sugar
1 pt. vegetable oil (can use a bit less if you want)
3 Tbsp. salt
2 tsps. celery seed
1 tsp. chipotle powder
1 tsp. curry powder
2 tsp. fennel seed
1 tsp. dill weed
one 4.5 oz jar minced or finely chopped garlic

1. Wash peppers & slice in rings (wear plastic gloves for this). Leave in seeds if you want them really hot. Brown, drain & crumble up sausage. In large kettle, combine ketchup, chili sauce, vinegar, sugar, oil, all the spices (crush the fennel first) and chopped garlic. Bring to a boil.

2. Add peppers, onions, and sausage; return to a boil. Cook on medium heat for about 10 minutes.

3. Pack into hot sterilized jars (leave about ½ in. head space for expansion). Place jars in a hot water bath for 10-15 minutes. Move to draft-free spot & wait for jars to seal. After 24 hrs., tighten down the rings and wash & dry jars for storage.

Note:
If desired, instead of the Italian sausage, you can cut up 2-3 lbs. of hot dogs or red-hot sausages....up to you!

West Virginia State Fire Department
Firefighter Carol Nolte, West Virginia

"If the smoke is moving faster than you can, you have been warned!"
"*The Art of Reading Smoke*"

BRAD'S CHICKEN POT PIE

Appleton is a city in Outagamie, Calumet, and Winnebago counties in the U.S. state of Wisconsin. One of the Fox Cities, it is situated on the Fox River, 30 miles (48 km) southwest of Green Bay and 100 miles (161 km) north of Milwaukee.

2 (10 oz) cans Campbell's Cream of chicken soup
2 (about 9 oz) pkgs. frozen mixed vegetables, thawed
2 C. cubed, cooked chicken
1 large potato, cubed and nuked until just soft
1 C. chicken broth combined with 2 Tbsp. flour
2 Pillsbury pie shells
salt, pepper, dash of cayenne

1. Preheat oven to 400°F.

2. Mix soup, vegetables, chicken, seasonings and broth. (The mixture should be sort of thick.) Place in pie shell.

3. Cover with second shell, fold and pinch shells together.

4. Cut 3 slits across top shell to let steam escape.

5. Bake for 35 minutes until golden. (Cover pie edge if it begins to brown too quickly).

The first pumper using a single engine for pumping and propulsion was manufactured in the United States in 1907.

Station # 2's Famous Rhubarb Tort

The Appleton Fire Department, located in Appleton, Wisconsin, has a staff of 96 men and women providing service from six fire stations strategically located throughout the City of Appleton.

Crust:
¾ C. real butter softened
1½ C. flour
7½ Tbsp. powdered sugar
pat into pan bake at 325 for 15 min.

Filling (mix in bowl):
4 eggs
½ C. flour
4 C. rhubarb
2½ C. sugar
pinch salt
½ tsp. vanilla

1. Mix and let sit in bowl for 15 min. usually while the crust is baking.

2. Bake approximately 1 hour.

3. Best to let cool. Great with ice cream.

A Chili cookoff done right!

In France, fire protection is administered in sectors, except in Paris, where the fire department is operated by the Sapeurs-Pompiers, a brigade of the French army, and in Marseille, where it is administered by the navy.

Cheese Bread Bubbles

The Riverton Volunteer Fire Department was organized in June of 1906 when a hand pulled hose wagon was officially put into service. The Department purchased a chemical engine on a hand cart in 1909 and sometime later added a ladder cart.

1 lb. loaf frozen white bread dough
2½ oz jar sliced dried beef
6 Tbsp. butter, melted
32 (½") cubes of cheese, your choice
½ tsp. garlic powder

1. Let dough thaw slightly.

2. Divide dough in half, and then cut each half into 16 pieces. Cut 16 slices of beef in half.

3. Fold beef slice in half lengthwise and wrap around a cheese cube. Shape each dough piece into a ball around a cube of beef-cheese; seal well.

4. Mix butter and garlic powder. Roll each ball in butter mixture and place in a greased Bundt pan.

5. Pour any leftover butter over dough balls.

6. Cover with towel and let rise in warm place until doubled, about 1½ hours. Bake at 375 degrees for 30 minutes.

7. Invert onto plate. Pull balls apart; serve hot.

"If you thought it was hard getting into the job--wait until you have to hang the "fire gear" up and walk away!"
Harry Lauder

Company Casserole

The Yellowstone fires of 1988 together formed the largest wildfire in the recorded history of the U.S.'s Yellowstone National Park. Starting as many smaller individual fires, the flames spread quickly out of control with increasing winds and drought and combined into one large conflagration, which burned for several months. The fires almost destroyed two major visitor destinations and, on September 8, 1988, the entire park was closed to all non-emergency personnel for the first time in its history.

8 oz pkg. wide noodles, cooked and drained
1½ lbs. ground beef
¼ tsp. garlic salt
16-oz can tomato sauce

To the Noodles:
1 C. cottage cheese
1 C. sour cream
6 green onions, chopped

1. Brown and drain the ground beef. Season with garlic salt. Add the tomato sauce. Simmer for 5 minutes

2. In a 9x13" pan, layer half the meat, all of the noodle/cheese mixture, then the rest of the meat in the pan. (It also works if you just mix it all together.)

3. Sprinkle 1 cup grated cheddar cheese over the top. Bake at 350 degrees for 30 minutes, uncovered.

Variation: Add salsa instead of the tomato sauce.

Firefighters and their toys, lined up for inspection

Riverton Volunteer Fire Department
Fire Chef Anne Metzler, Riverton, Wyoming

Diplomacy - the art of telling someone to go to hell, and having them look forward to the trip.

Five-Cup Salad

The city is an incorporated entity of the state of Wyoming, on land ceded from the reservation in 1906, a situation that often makes it subject to jurisdictional claims by the nearby Eastern Shoshone and Northern Arapaho tribes. The community was named Riverton because of the four rivers that meet there.

1 C. miniature marshmallows
1 C. mandarin oranges, drained
1 C. crushed or chunk pineapple, drained
1 C. sour cream
1 C. coconut

1. Combine all ingredients.

2. Chill at least 1 hour or overnight. Makes 6-8 servings.

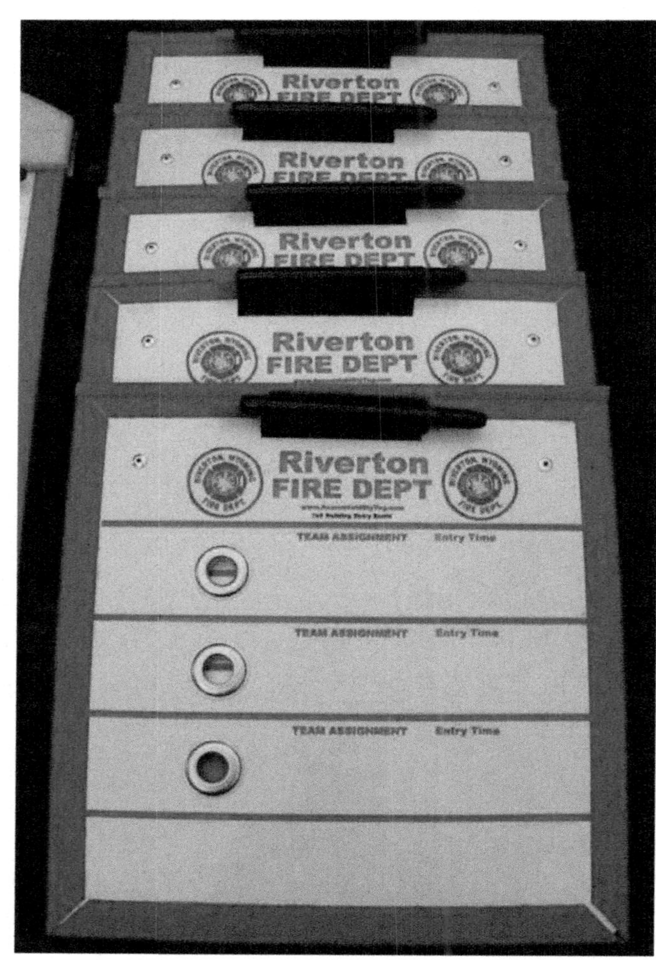

Invented by an experienced fire chief, this unique mobile command system is critical to incident accountability. The Command Board system provides immediate information about resource location, arrivals, teams in the deck, teams in rehab, who is in the hot zone, and much more.

TOWER 1